W9-APL-795

Introduction to Analysis

THIRD EDITION

Introduction to Analysis

THIRD EDITION

Edward D. Gaughan

Brooks/Cole Publishing Company
Pacific Grove, California

Consulting Editor: *Robert J. Wisner*

Brooks/Cole Publishing Company
A Division of Wadsworth, Inc.

Printed in the United States of America

10 9 8 7 6 5 4 3 2

Library of Congress Cataloging-in-Publication Data

Gaughan, Edward.
Introduction to analysis.

Includes index.
1. Mathematical analysis. I. Title.
QA300.G34 1987 515 86-28323
ISBN 0-534-07602-5

Sponsoring Editor: *Jeremy Hayhurst*
Editorial Assistant: *Amy Mayfield*
Production Editor: *Joan Marsh*
Manuscript Editor: *Patricia Cain*
Interior and Cover Design: *Sharon L. Kinghan*
Art Coordinator: *Sue C. Howard*
Interior Illustration: *Scientific Illustrators*
Typesetting: *Omegatype Typography*

Preface

Introduction to Analysis is designed to bridge the gap between the intuitive calculus normally offered at the undergraduate level and the sophisticated analysis courses the student reaches in the senior or first-year-graduate level. Through a rigorous approach to the usual topics handled in one-dimensional calculus—limits, continuity, differentiation, integration, and infinite series—the book offers a deeper understanding of the ideas encountered in calculus. Although the text assumes that the reader has completed several semesters of calculus, this assumption is necessary only for some of the motivation (of theorems) and examples.

The book has been written with two important goals in mind for its readers: the development of a rigorous foundation for the basic topics of analysis and the less tangible acquisition of an accurate intuitive feeling for analysis. In the interests of these goals, considerable time is devoted to motivating and developing new concepts. Economy of space is often sacrificed so that ideas car be introduced in a natural fashion.

This 3rd edition contains changes recommended by the users of earlier editions of the book. Chapter 0 contains introductory material on sets, functions, relations, mathematical induction, recursion, equivalent and countable sets, and the set of real numbers. The set of real numbers is postulated as an ordered field with the least upper bound property as in the 2nd edition. The section on mathematical induction and recursion is new to this edition in response to suggestions by the reviewers.

Chapters 1 through 4 remain essentially unchanged in content and contain the material on sequences, limits of functions, continuity, and differentiation. Chapter 5 is devoted to the Riemann integral, rather than the Riemann-Stieltjes integral treated in the first edition. Chapter 6 treats infinite series, and Chapter 7 contains material on sequences and series of functions.

The exercise sets have been expanded significantly, and offer a selection of exercises with level of difficulty ranging from very routine to quite challenging. The starred exercises are of particular importance, because they contain facts vital to the development of later sections. The star is *not* used to indicate the more difficult exercises.

At the end of each chapter, you will find a PROJECT. This is a set of exercises designed to lead you through the proof of a result that is closely related to the content of the chapter. A PROJECT is distinguished from an exercise in that the PROJECT involves three or more steps in the proof and for the beginner, such proofs are very difficult. You will notice that in later chapters, more and more is left to the reader in the development of the PROJECTS.

In the course of this exposition, a number of famous names are mentioned: Dedekind, Cauchy, Bolzano, Weierstrass, Riemann, and others. A serious student should complement his mathematical knowledge by seeking to know something about the men who have made important contributions to analysis. The reader is urged to indulge in a little historical research when encountering the names of these people.

I would like to acknowledge the reviewers: Professors William Ballard, University of Montana; Terry Herdman, Virginia Polytechnic Institute; Charles Himmelberg, University of Kansas; L. G. Hoye, University of Lethbridge; Daniel Kocan, State University of New York, Potsdam; Walter M. Patterson, III, Lander College; Linda Sons, Northern Illinois University; and J. L. Williams, University of Sydney.

My sincere appreciation is also extended to: Robert J. Wisner, who, some 20 years ago, urged me to write the first edition of this book, and has made many helpful suggestions relative to each of the three editions; Jack N. Thornton, who as managing editor of Brooks/Cole Publishing Co. accepted and nurtured a fledgling author through both the first and second editions; and to Jeremy Hayhurst, mathematics editor at Brooks/Cole Publishing Co., for his support and encouragement in the preparation of this third edition. Of course, this book would not be in your hands without the excellent work of Joan Marsh who carried it and me through the stages of production. My thanks go to her also.

Edward D. Gaughan

Contents

Introduction to Analysis

THIRD EDITION

Preliminaries

Before attempting to study analysis, one must be able to read and communicate mathematics intelligently. This fact is not unique to analysis but is true in all of mathematics. This chapter presents some of the basic vocabulary of mathematics; in fact, its contents, with some rearrangements, may be similar to the beginning chapter of a book at this level in algebra, topology, or other topics in mathematics. This similarity is not accidental. A certain basic vocabulary is common to a good share of mathematics. The chapter does, however, exclude anything unnecessary for the assimilation of the material to come in this book.

A few words of both warning and encouragement are in order. First, you should realize that only the fundamentals of your mathematical vocabulary are presented here; the proper usage comes with practice and increasing mathematical maturity. Thus, at the beginning it may seem a bit awkward to use new and possibly unfamiliar ideas in the development of additional concepts. Now the words of encouragement: The initial ideas presented and the theorems proved are quite simple; hence, they will give you many chances to practice your vocabulary in settings where intuition can help to guide your thinking. You are encouraged to play this game quite seriously by giving precise proofs to easy theorems, and thus gaining practice in clear and precise mathematical expression—an ability that will be invaluable as the material becomes more difficult in later chapters.

One last bit of admonishment is appropriate. Mathematics, by its very nature, begs to be communicated. It is difficult to imagine a mathematician who, upon discovering a new fact or proof, does not have a burning desire to shout it from the rooftops. In fact, any professional mathematician—teacher, researcher, or what have you—must be able to communicate with others. Those who say that mathematics is completely incomprehensible have either failed to learn the language of mathematics or have had the misfortune of trying to learn mathematics from someone who cannot or will not use the language properly. Much symbolism is used in mathematics, but each symbol or set of symbols must stand for a word or phrase in the language used, which is English in this case. In particular, the sentences formed with symbols must make sense when translated into words and must impart the meaning intended. A good test for your use of symbols is to have someone read your writing with a critical eye, translating it aloud to you. Then you can see whether it meets the test for clarity and meaning.

0.1 SETS

We shall begin with the naive notion of a *set*. A very simple approach will be sufficient for our purposes in this book. We shall think of a set as a collection of objects. Note that the word *collection* is as undefined in this setting as is the word *set*. In order to build the intuitive idea of a set, consider the following examples:

1. The set of all natural numbers.
2. The set of all the letters of the Greek alphabet.
3. The set of all rational roots of the equation $x^2 + 1 = 0$.
4. The set consisting of the rational numbers 1, 2, and 3.
5. The set of all integers less than 4 and greater than 0.

Note that in examples 1, 2, 3, and 5 the set was described by a rule for determining which objects belong to the set, whereas in example 4 the elements were explicitly named. In particular, observe that those objects that pass the test for membership in the set described in example 5 are precisely the objects listed in example 4. It would be quite disconcerting if these two sets were different—that is, if a set were determined by the method used to describe it rather than by the objects belonging to it. That this will not be the case is made clear by the following definition.

DEFINITION If A and B are sets, then $A = B$ if and only if every object belonging to A also belongs to B and every object belonging to B also belongs to A.

This definition makes it obvious that the sets described in examples 4 and 5 are equal. To show that two sets A and B are not equal, it suffices to find an object that belongs to one set and not to the other.

DEFINITION If A is a set and x is an object that belongs to A, we say "x is an *element* of A" or "x is a *member* of A" and write $x \in A$. If x does not belong to A, we write $x \notin A$.

Let us use this notation to rewrite the definition of equality for sets. If A and B are sets, then $A = B$ if and only if for each object x, $x \in A$ implies $x \in B$, and $x \in B$ implies $x \in A$. Let A_1 be the set in example 1, A_2 be the set in example 2, and so on. Now, as observed above, $A_4 = A_5$ and $A_1 \neq A_4$, because $13 \in A_1$ and $13 \notin A_4$. Note, however, that each member of A_4 is also a member of A_1. In a sense, A_4 is a part of A_1.

DEFINITION If A and B are sets such that each member of A is also a member of B, then A is a *subset* of B or A is *contained* in B, written $A \subset B$. If $A \subset B$ and $A \neq B$, then A is a *proper* subset of B.

We are now in a position to state and prove the first theorem of this book. It will be useful later when we wish to prove that two sets are equal. Read and reread the statement of the theorem and reflect on its meaning. Of course, in light of the three definitions just given, it seems obvious that the theorem is true. But, let us use the opportunity to present an easy proof and discuss the method of proof to be presented.

This format is typical for a theorem in mathematics; it is an "if and only if" theorem. Basically, the theorem has two parts: (1) if $A = B$, then $A \subset B$ and $B \subset A$ and (2) if $A \subset B$ and $B \subset A$, then $A = B$. To prove the theorem, we must prove that *both* (1) and (2) hold. Follow along.

0.1 THEOREM Let A and B be sets. Then $A = B$ if and only if $A \subset B$ and $B \subset A$.

Proof. Assume $A = B$. Then, if $x \in A$, then $x \in B$ since $A = B$ (see the definition of $A = B$); hence $A \subset B$. Likewise, if $x \in B$, then $x \in A$ since $A = B$, and we have $B \subset A$. Thus, if $A = B$, then $A \subset B$ and $B \subset A$.

Assume now that $A \subset B$ and $B \subset A$. We want to show that $A = B$. Choose any $x \in A$. Since $A \subset B$, then $x \in B$. If $x \in B$, then since $B \subset A$, we have $x \in A$. Thus, by the definition of set equality, $A = B$. This concludes the proof.

It is convenient to have a variety of ways of describing sets. We have already seen sets defined by a rule for membership and by a list of the members of the set. If $P(x)$ is a statement concerning an object x, then we denote $\{x : P(x)\}$ the set of all objects x such that $P(x)$ is true. The kind of statements that are permissible here is a subject beyond the scope of this book. It suffices to declare that the types encountered in this book are permissible.

■ **Example 0.1** The set of all natural numbers is $\{x: x \text{ is a natural number}\}$. The set of all integers less than 4 and greater than 0 is equal to $\{n : 0 < n < 4 \text{ and } n \text{ is an integer}\}$. ■

There are some sets to which we give special names, and we list them below:

1. J = set of all natural numbers—i.e., the set of all positive integers.
2. Z = set of all integers.
3. R = the set of all real numbers.
4. For $a < b$, $[a, b] = \{x : x \in R \text{ and } a \leq x \leq b\}$.
5. For $a < b$, $[a, b) = \{x : x \in R \text{ and } a \leq x < b\}$.
6. For $a < b$, $(a, b] = x : x \in R \text{ and } a < x \leq b\}$.
7. For $a < b$, $(a, b) = \{x : x \in R \text{ and } a < x < b\}$.

8. We use the symbol $\varnothing$ for the empty set; that is $\varnothing$ is the set that contains *no* elements.

DEFINITION If A and B are sets, the *union* of A and B, written $A \cup B$, is defined to be the set of all objects that belong to either A or B or possibly both. In the notation used previously, $A \cup B = \{x : x \in A \text{ or } x \in B\}$.

DEFINITION If A and B are sets, the *intersection* of A and B, written $A \cap B$, is defined to be the set of all objects that belong to both A and B. In other words, $A \cap B = \{x : x \in A \text{ and } x \in B\}$.

To partially digest these new definitions, let us consider a few more examples. Let

$$A = \{\alpha, \beta, \gamma\}, \quad B = \{1, 2, 4, \alpha, \beta\}, \quad \text{and} \quad C = \{1, 2, 3, 4, 5\}.$$

Now $A \cap B = \{\alpha, \beta\}$, $A \cup B = \{1, 2, 4, \alpha, \beta, \gamma\}$, and $A \cap C$ is empty. If two sets A and B have the property that $A \cap B$ is empty, we say that they are *disjoint*.

An appropriate theorem to state at this stage contains some parts that are quite obvious from the definition of union and intersection and some that are not quite so obvious.

0.2 THEOREM Let A, B, and C be sets. Then

i. $A \cap B = B \cap A$.
ii. $A \cup B = B \cup A$.
iii. $(A \cap B) \cap C = A \cap (B \cap C)$.
iv. $(A \cup B) \cup C = A \cup (B \cup C)$.
v. $A \cap (B \cup C) = (A \cap B) \cup (A \cap C)$.
vi. $A \cup (B \cap C) = (A \cup B) \cap (A \cup C)$.

Parts (i) through (iv) have very obvious proofs, that will not be supplied here. The proof of (v) follows, and that for (vi) is left as Exercise 3.

Proof. In proving (v) we shall use Theorem 0.1 to show that

$$(A \cap (B \cup C) \subset (A \cap B) \cup (A \cap C)$$

and

$$(A \cap B) \cup (A \cap C) \subset A \cap (B \cup C),$$

and hence the two sets are equal.

Suppose $x \in A \cap (B \cup C)$. Then $x \in A$ and $x \in B \cup C$. If $x \in B$, then since $x \in A$, $x \in A \cap B$. If $x \not\in B$, then since $x \in B \cup C$, $x \in C$ and also $x \in A$; hence, $x \in A \cap C$. Thus, in particular, $x \in A \cap B$ or $x \in A \cap C$. In other words,

$$x \in (A \cap B) \cup (A \cap C).$$

We have shown that $A \cap (B \cup C) \subset (A \cap B) \cup (A \cap C)$.
Suppose

$$x \in (A \cap B) \cup (A \cap C).$$

Then $x \in A \cap B$ or $x \in A \cap C$. If $x \in A \cap B$, then $x \in A$ and $x \in B$; hence, $x \in A$ and $x \in B \cup C$ since $B \subset B \cup C$. Thus,

$$x \in A \cap (B \cup C).$$

If $x \notin A \cap B$, then $x \in A \cap C$; hence, $x \in A$ and $x \in C$, but $C \subset B \cup C$, so that $x \in A \cap (B \cup C)$. We have shown

$$(A \cap B) \cup (A \cap C) \subset A \cap (B \cup C),$$

and the proof is complete.

Some of the reasonably easy relations concerning sets given in the exercises will be used later, and you are urged to prove them for your own practice—for the simple reason that a proof makes a theorem much easier to remember.

DEFINITION Let A and B be sets. Then the *complement* of A *relative* to B, written $B \setminus A$, is defined to be the set of all objects belonging to B but not to A. In other words,

$$B \setminus A = \{x : x \in B \text{ and } x \notin A\}.$$

Let $A = \{1, 2, 3, 4, 5, 6\}$ and $B = \{2, 4, 6, 8, 10\}$. Then $B \setminus A = \{8, 10\}$ and $A \setminus B = \{1, 3, 5\}$. The following form of what are called *De Morgan's Laws* may look a bit cumbersome because of our decision to speak of the complement of A relative to B rather than the complement of a set relative to some implied universal set. However, this path leaves little room for misunderstanding.

0.3 THEOREM Let A, B, and C be sets. Then

i. $A \setminus (B \cap C) = (A \setminus B) \cup (A \setminus C)$,
ii. $A \setminus (B \cup C) = (A \setminus B) \cap (A \setminus C)$.

Proof. Again, the method of proof will be to use Theorem 0.1, showing that the set on the left side of the proposed equality is a subset of the set on the right and vice versa. Although this method of proof will be used quite often in such theorems, it will not always be advertised in the future.

(i) Let

$$x \in A \setminus (B \cap C).$$

Then $x \in A$ and $x \notin B \cap C$. Since $x \notin B \cap C$, either $x \notin B$ or $x \notin C$. If $x \notin B$, then, since $x \in A$, $x \in A \setminus B$. If $x \notin C$, then since $x \in A$, $x \in A \setminus C$. Hence,

$$x \in (A \setminus B) \cup (A \setminus C).$$

Thus,

$$A \setminus (B \cap C) \subset (A \setminus B) \cup (A \setminus C).$$

Suppose now that

$$x \in (A \setminus B) \cup (A \setminus C).$$

Then either $x \in A \setminus B$ or $x \in A \setminus C$. If $x \in A \setminus B$, then $x \in A$ and $x \notin B$, so that $x \notin B \cap C$; thus, $x \in A \setminus (B \cap C)$. If $x \in A \setminus C$, then $x \in A$ and $x \notin C$, so that $x \notin B \cap C$; thus

$$x \in A \setminus (B \cap C).$$

We have shown that $(A \setminus B) \cup (A \setminus C) \subset A \setminus (B \cap C)$. This result coupled with that of the preceding paragraph completes the proof of (i).

The proof of (ii) is left as Exercise 4.

In order to gain a deeper understanding of Theorem 0.3, let us consider a special case and invent some new notation. Suppose B and C are both subsets of A. If S is a subset of A, let S^* denote the complement of S relative to A so that $S^* = A \setminus S$. In this setting, the theorem states that

$$(B \cap C)^* = B^* \cup C^* \quad \text{and} \quad (B \cup C)^* = B^* \cap C^*.$$

In cruder terms, the complement of the intersection is the union of the complements, and the complement of the union is the intersection of the complements.

The notions of union and intersection can be generalized to unions and intersections of more than two sets. But first some notation is needed.

DEFINITION Let Λ be a set, and suppose for each $\lambda \in \Lambda$, a subset A_λ of a given set S is specified. The collection of sets A_λ is called an *indexed family* of subsets of S with Λ as the index set. We denote this by $\{A_\lambda\}_{\lambda \in \Lambda}$.

DEFINITION If $\{A_\lambda\}_{\lambda \in \Lambda}$ is an indexed family of sets, define

$$\bigcap_{\lambda \in \Lambda} A_\lambda = \{x : x \in A_\lambda \text{ for all } \lambda \in \Lambda\}$$

and

$$\bigcup_{\lambda \in \Lambda} A_\lambda = \{x : x \in A_\lambda \text{ for some } \lambda \in \Lambda\}$$

The reader is invited to generalize some of the previous theorems in light of these new ideas.

There is one logical difficulty here that needs to be pointed out. If Λ is the empty set, then it is easy to see that $\bigcup_{\lambda \in \Lambda} A_\lambda$ is empty; however, it is not clear what to expect of $\bigcap_{\lambda \in \Lambda} A_\lambda$. This could be overcome by insisting that all index sets be nonempty, and this shall be done in some cases. In a context where all sets considered are understood to be subsets of a given set S, the common usage is to let

$$S = \bigcap_{\lambda \in \varnothing} A_\lambda$$

where $\varnothing$ is the empty set. The reason for this discussion is to alert the reader of possible problems that might arise from this situation, not to give a solution.

In the event that the index set is the set J of positive integers or a finite subset thereof, some special notation may be adopted. For example, $\bigcup_{n \in J} A_n$ may also be written $\bigcup_{n=1}^{\infty} A_n$, and if $S = \{1, 2, \ldots, k\}$, we may write $\bigcup_{n=1}^{k} A_n$ instead of $\bigcup_{n \in S} A_n$. Of course, similar notation may be used for the intersection of a family of sets.

■ **Example 0.2** Let $A_n = \left(0, \frac{1}{n}\right)$ for each $n \in J$. Then

$$\bigcap_{n=1}^{\infty} A_n = \varnothing \quad \text{(the empty set).}$$

Let us see why. If $x \in \bigcap_{n=1}^{\infty} A_n$ then $x \in \left(0, \frac{1}{n}\right)$ for each $n \in J$; that is, $0 < x < \frac{1}{n}$ for all $n \in J$. However, if we pick n_0 to be any positive integer greater than $\frac{1}{x}$, then $0 < \frac{1}{n_0} < x$ and $x \notin \left(0, \frac{1}{n_0}\right)$, contrary to our assumption. Thus $\bigcap_{n=1}^{\infty} A_n$ is the empty set since no real number can satisfy the conditions for membership. We are assuming some prior knowledge of real numbers, integers, and inequalities, but we will discuss these details later in this chapter. ■

■ **Example 0.3** Let $B_n = \left(\frac{1}{n}, 1\right)$ for each $n \in J$. Then

$$\bigcup_{n=1}^{\infty} B_n = (0, 1).$$

Again, let us see why. Suppose $0 < x < 1$. There is a positive integer m such that $\frac{1}{x} < m$, hence $\frac{1}{m} < x < 1$; that is $x \in \left(\frac{1}{m}, 1\right)$. But then, since x belongs to one set in the indexed family, it must belong to the union of the sets in the indexed family. So $(0, 1) \subset \bigcup_{n=1}^{\infty} B_n$. On the other hand, if $x \in \bigcup_{n=1}^{\infty} B_n$, then $x \in B_n$ for some $n \in J$; hence, $0 < \frac{1}{n} < x < 1$. In any case $0 < x < 1$. Thus $\bigcup_{n=1}^{\infty} B_n \subset (0, 1)$. Therefore $\bigcup_{n=1}^{\infty} B_n = (0, 1)$. ■

A generalization of De Morgan's Laws can be stated and proved for indexed families of sets, and such a generalization will be useful later.

0.4 THEOREM (De Morgan's Laws) Let S be a set and $\{A_\lambda\}_{\lambda\in\Lambda}$ be an indexed family of subsets of S. Then

i. $S\setminus(\bigcup_{\lambda\in\Lambda} A_\lambda) = \bigcap_{\lambda\in\Lambda}(S\setminus A_\lambda)$.
ii. $S\setminus(\bigcap_{\lambda\in\Lambda} A_\lambda) = \bigcup_{\lambda\in\Lambda}(S\setminus A_\lambda)$.

Proof. We will prove (i) and leave (ii) as Exercise 11. Assume $x \in S\setminus(\bigcup_{\lambda\in\Lambda} A_\lambda)$. Then $x \in S$ and $x \notin \bigcup_{\lambda\in\Lambda} A_\lambda$. Therefore $x \in S$ and $x \notin A_\lambda$ for all $\lambda \in \Lambda$. Therefore, $x \in S\setminus A_\lambda$ for all $\lambda\in\Lambda$. This means that $x \in \bigcap_{\lambda\in\Lambda}(S\setminus A_\lambda)$. On the other hand, if $x \in \bigcap_{\lambda\in\Lambda}(S\setminus A_\lambda)$, then $x \in S\setminus A_\lambda$ for each $\lambda\in\Lambda$. Thus $x\in S$ and $x \notin A_\lambda$ for each $\lambda\in\Lambda$. This implies that $x\in S$ and $x\notin \bigcup_{\lambda\in\Lambda} A_\lambda$. So $x \in S\setminus(\bigcup_{\lambda\in\Lambda} A_\lambda)$. We have shown that

$$S\setminus\Big(\bigcup_{\lambda\in\Lambda} A_\lambda\Big) \subset \bigcap_{\lambda\in\Lambda}(S\setminus A_\lambda)$$

and

$$\bigcap_{\lambda\in\Lambda}(S\setminus A_\lambda) \subset S\setminus\Big(\bigcup_{\lambda\in\Lambda} A_\lambda\Big)$$

Therefore the two sets are equal.

■ **Example 0.4** Using Example 0.2 and De Morgan's Laws, we see that

$$\bigcup_{n=1}^{\infty}\left(R\setminus\left(0,\frac{1}{n}\right)\right) = R\setminus\bigcap_{n=1}^{\infty}\left(0,\frac{1}{n}\right) = R\setminus\varnothing = R. \quad ■$$

0.2 RELATIONS AND FUNCTIONS

Given objects x and y, $x \neq y$, it is easy to form the set whose only members are x and y: We can write the set as either $\{x, y\}$ or $\{y, x\}$ since the method of description doesn't determine the set, but determines only the objects that belong to the set. We are interested in another object one may construct from two objects, namely an *ordered pair.* The distinction we desire here is that each of the objects has a special place—as opposed to the set, in which order is unimportant. This concept is not unfamiliar, for the notion of an ordered pair of real numbers is encountered in algebra at a tender age-for example, as in the case of fractions wherein numerators and denominators are distinguished.

DEFINITION The *ordered pair* (x, y) is defined to be the set $\{\{x\}, \{x, y\}\}$.

Give this definition a little time to soak in. The ordered pair (x, y) is the set whose members are the set whose only member is x and the set whose members are x and y. This is our first encounter here with a set that has other sets as members. If $x = y$, then $(x, x) = \{\{x\}\}$. Should this definition seem disconcerting, remember that its purpose is to single out one object as the "first coordinate" and another as the "second coordinate." After proving the following theorem, we shall see that the definition serves the desired purpose.

0.5 THEOREM $(x, y) = (u, v)$ if and only if $x = u$ and $y = v$.

Proof. If $x = u$ and $y = v$, then

$$(x, y) = \{\{x\}, \{x, y\}\} = \{\{u\}, \{u, v\}\} = (u, v).$$

Suppose now that $(x, y) = (u, v)$. Then $\{\{x\}, \{x, y\}\} = \{\{u\}, \{u, v\}\}$. If $x = y$, then

$$\{\{x\}\} = \{\{u\}, \{u, v\}\}.$$

Since the set on the left has only one member, the same must be true of the set on the right. This can be true only if $\{u\} = \{u, v\}$, which in turn can be true only if $u = v$. Thus, we have

$$\{\{x\}\} = \{\{u\}\},$$

and so $\{x\} = \{u\}$, $x = u$, and, of course $y = x = u = v$. Now suppose x and y are distinct. Then $\{x\}$ and $\{x, y\}$ are distinct sets and are the only members of $\{\{x\}, \{x, y\}\}$. By the assumed equality,

$$\{x\} \in \{\{u\}, \{u, v\}\}.$$

The case $\{x\} = \{u, v\}$ is impossible since u and v must be distinct, and thus $\{u, v\}$ has two elements, whereas $\{x\}$ has only one element. Thus $\{x\} = \{u\}$, so $x = u$. Also

$$\{x, y\} \in \{\{u\}, \{u, v\}\},$$

and the only remaining element that can be equal to $\{x, y\}$ is $\{u, v\}$. But $x = u$, and by the equality $\{x, y\} = \{u, v\}$, we must have $y = v$. This concludes the proof.

The concept of an ordered pair sets the stage for another way of combining two sets to form a new set. In analytic geometry, one considers the set of all ordered pairs of real numbers—a new set constructed from an old one; a generalization of this will be the next definition.

DEFINITION If A and B are sets, define the *Cartesian product* of A and B, written $A \times B$, to be the sets of all ordered pairs (a, b) such that $a \in A$ and $b \in B$. Thus $A \times B = \{(a, b) : a \in A \text{ and } b \in B\}$.

■ **Example 0.5** As an example, let $A = \{1, 2, 3\}$ and $B = \{2, 4, 6\}$. Then

$$A \times B = \{1, 2), (1, 4), (1, 6), (2, 2), (2, 4), (2, 6), (3, 2), (3, 4), (3, 6)\}. \quad ■$$

Note that, in general, if A has m elements and B has n elements, then $A \times B$ has mn elements.

DEFINITION A *relation* is a set of ordered pairs. A *function* is a relation F such that if $(x, y) \in F$ and $(x, z) \in F$, then $y = z$. If F is a function, the *domain* of F, written dom F, is defined to be

$$\text{dom } F = \{x : (x, y) \in F \text{ for some } y\},$$

and the *image* of F, written im F, is defined to be

$$\text{im } F = \{y : (x, y) \in F \text{ for some } x \in \text{dom } F\}.$$

Examine the definition of function and reflect on the meaning. For each $x \in \text{dom } F$, there is *exactly* one element $y \in \text{im } F$ such that $(x, y) \in F$. It is customary to refer to such a y as $F(x)$. Now observe how F works. Choose any $x \in \text{dom } F$; then F produces a unique element y in im F. We call it $F(x)$. It is useful to think of F in a dynamic way. Input x and the function F produces an output, $F(x)$. Often $F(x)$ is computed by a formula or rule.

Suppose that F is a function with dom $F = A$ and im $F \subset B$. We would refer to B as the *codomain* of F and write this symbolically as:

$$F: A \to B.$$

One reason for specifying the codomain rather than the image of F is that determining the image of F is sometimes a very difficult problem.

■ **Example 0.6** Functions may be defined in a variety of ways. A popular means for defining a function is to give a rule as well as the domain of the function. For example, define a function $f: J \to J$ as follows:

$$f(n) = 2n - 1 \quad \text{for each } n \in J.$$

Notice that the rule is unambiguous; that is, given $n \in J$, there is no doubt about $f(n)$. Notice that the domain is specified. ■

Consider how the functions $f(x) = x^3$, $x \in R$, and $g(x) = \sqrt[3]{x}$, $x \in R$. There is a special relationship between f and g that we will discuss shortly. For the moment, notice that $f(2) = 8$ and $g(8) = 2$. This means that g "undoes" what f "does" and vice versa. You have seen other pairs of functions with this property before, and we examine the idea in detail.

DEFINITION Let S be a relation. Then the *converse* of S, written $\hat{S}$ is defined by

$$\hat{S} = \{(x, y) : (y, x) \in S\}.$$

It would be interesting to discover under what conditions the converse of a function is also a function. Suppose F is a function with (x, y) and (x, z) both belonging to $\hat{F}$. In order that $\hat{F}$ be a function, it must be true that $y = z$. Translating back to F, for all (y, x) and (z, x) in F, in must be true that $y = z$; or, in other words, if $F(y) = F(z)$, then $y = z$. This idea is important enough to deserve a special definition.

DEFINITION A function f is 1–1 ("pronounced one-to-one") if and only if, for all y, z in the domain of f, $f(y) = f(z)$ implies $y = z$. In essence, this says that a 1–1 function is one that assumes each value in its image exactly once.

Many theorems in mathematics take the following form: Property A is satisfied if and only if Property B is satisfied. It has become accepted practice in mathematics literature to use "iff" as an abbreviation for "if and only if." Thus, the theorem would be written: Property A is satisfied iff Property B is satisfied.

0.6 THEOREM Let F be a function. Then $\hat{F}$ is a function iff F is a 1–1 function. If $\hat{F}$ is a function, then dom $\hat{F}$ = im F and im $\hat{F}$ = dom F.

Proof. Assume $\hat{F}$ is a function. Suppose x and y belong to the domain of F with $F(x) = F(y) = z$. Then $(x, z) \in F$ and $(y, z) \in F$ so that $(z, y) \in \hat{F}$ and $(z, x) \in \hat{F}$. Since $\hat{F}$ is a function $y = x$. Thus, if $\hat{F}$ is a function, then F is 1–1.

Now assume F is 1–1. Let $(u, w) \in \hat{F}$ and $(u, v) \in \hat{F}$. To show that $\hat{F}$ is a function, we must show that $w = v$. By the definition of $\hat{F}$, $(w, u) \in F$ and $(v, u) \in F$, making $u = F(w) = F(v)$. By assumption, F is 1–1, hence $w = v$. The facts that im F = dom $\hat{F}$ and dom F = im $\hat{F}$ are immediate.

If F is a 1–1 function, we shall write F^{-1} in place of $\hat{F}$ and call F^{-1} the *inverse* of F. Furthermore, if F is 1–1, then $\hat{\hat{F}} = F$ is a function, so $\hat{F} = F^{-1}$ is also 1–1.

Suppose f: $A \to B$ and $T \subset A$. Define

$$f(T) = \{f(t) : t \in T\}.$$

$f(T)$ is called the *image* of T under f. In particular $f(A) = \text{im} f$. If $C \subset B$, define

$$f^{-1}(C) = \{a \in A : f(a) \in C\}.$$

$f^{-1}(C)$ is called the *inverse image* of C under f. Note that this definition is given without any assumption that f is 1–1, so it is not assumed here that $\hat{f}$ is a function.

If f: $A \to B$ is a 1–1 function, f is sometimes called an *injection* or an *injective* function. When im $f = B$, we say that f is a function from A *onto* B and call f a *surjection* or a *surjective* function. If f: $A \to B$ is both 1–1 and onto, f is called a *bijection* or a *bijective* function.

DEFINITION If f: $A \to B$ and g: $B \to C$, the *composition* of g by f, written $g \circ f$ is defined to be the set

$$\{(x, y) : \text{there is } w \in B \text{ such that } (x, w) \in f \text{ and } (w, y) \in g\}.$$

0.7 THEOREM If f: $A \to B$ and g: $B \to C$, then $g \circ f$ is a function and $g \circ f : A \to C$.

Proof. Suppose $(x, u) \in g \circ f$ and $(x, v) \in g \circ f$. We must show that $u = v$. By the definition of $g \circ f$ there are w_1 $w_2 \in B$ such that $(x, w_1) \in f$, $(w_1, u) \in g$, $(x, w_2) \in f$, and $(w_2, v) \in g$. Since f is a function, $w_1 = w_2$; since g is a function, $u = v$. Thus $g \circ f$ is a function.

Let f: $A \to B$ and g: $B \to C$. If $x \in A$, then there is $y \in B$ such that $(x, y) \in f$, or $y = f(x)$; also, there is $z \in C$ such that $(y, z) \in g$, or $z = g(y)$. This means that $(x, z) \in g \circ f$ and $z = g \circ f(x)$. But

$$z = g(y) = g(f(x)),$$

so

$$(g \circ f)(x) = g(f(x)).$$

and we see that the composition of g by f is accomplished by "following f by g. Also, it is clear the dom $(g \circ f) =$ dom f.

Though it is defined at length here, the composition of two functions is not a new idea. For example, the function $h(x) = \sin x^2$ is the composition of the function $g(x) = \sin x$ by $f(x) = x^2$, so that $h = g \circ f$. Note that $f \circ g$ is the function whose value at x is $(\sin x)^2$; hence $f \circ g \neq g \circ f$.

If S is a set, denote by 1_S the function from S onto S defined by $1_S(s) = s$ for each $s \in S$. Now, if f is a 1–1 function from A onto B, then we have seen that f^{-1} is a 1–1 function from B onto A. Upon examination of the definition of f^{-1}, it becomes clear that $f \circ f^{-1} = 1_B$ and $f^{-1} \circ f = 1_A$.

■ **Example 0.7** Let $S = \{x : x \in R \text{ and } x \geqslant 0\}$. Define f: $S \to R$ by $f(x) = x^2$. If $f(x) = f(y)$, then $x^2 = y^2$, which means that $x = y$ or $x = -y$. But, since both x and y must belong to S, then x and y are both zero or both have the same sign—i.e., both are positive. In either case $x = y$. Thus f is 1–1. Let us find f^{-1}. If $(x, y) \in f$ then $y = x^2$. Thus, if (y, x) belongs to f^{-1}, $y = x^2$, and because dom $f = S$, x must be the positive square root of y if $y \neq 0$ and x must be zero otherwise. Thus $f^{-1}(y) = \sqrt{y}$. We know that im $f = S$ (previous experience tells us that), so $f^{-1} : S \to S$. Notice that

$$f \circ f^{-1}(y) = f(f^{-1}(y)) = (\sqrt{y})^2 = y$$

and

$$f^{-1} \circ f(x) = f^{-1}(f(x)) = \sqrt{x^2} = x \quad \text{since } x \geqslant 0. \quad ■$$

■ **Example 0.8** Let $S = \{x : x \in R \text{ and } x \neq -1\}$. Define f: $S \to R$ by $f(x) = \dfrac{2x - 1}{x + 1}$. First of all, if $f(a) = f(b)$, then $\dfrac{2a - 1}{a + 1} = f(a) = f(b) = \dfrac{2b - 1}{b + 1}$. A bit of algebra shows that $a = b$, hence f is 1–1. Let us find im f and, in the process, we will discover f^{-1}. The number $y \in$ im f if and only if there is

$x \in \operatorname{dom} f = S$ such that $f(x) = y$—that is $y = \dfrac{2x - 1}{x + 1}$. Solving for x, we obtain $x = \dfrac{y + 1}{2 - y}$. This means that if $y \neq 2$ there is $x \in S$ such that $f(x) = y$. Thus $\operatorname{im} f = R \setminus \{2\}$. Also, this means that $f^{-1}(y) = \dfrac{y + 1}{2 - y}$. ∎

0.3 MATHEMATICAL INDUCTION AND RECURSION

In what is to follow, we will have need of a technique of proof called *mathematical induction*. In some cases, the use will be rather informal, as in the case of showing that every polynomial function has a limit at every point. On the other hand, in less obvious applications, such as Taylor's Theorem in Chapter 5, the approach will be formal and very precise.

Recursion will be used in the definition of certain functions and sequences defined on J. Usually a few initial values of the function are given and then the value at n is given in terms of the values of the function at k for some $k < n$.

The basis for mathematical induction is the well-ordering principle, which we assume without proof.

WELL-ORDERING PRINCIPLE Every nonempty subset of J has a smallest member.

0.8 THEOREM (PRINCIPLE OF MATHEMATICAL INDUCTION) If $P(n)$ is a statement for each $n \in J$ such that

a. $P(1)$ is true, and
b. for each $k \in J$, if $P(k)$ is true, then $P(k + 1)$ is true,

then $P(n)$ is true for all $n \in J$.

Reflect on the implication of Theorem 0.8. Basically it says that, if $P(1)$ is true, and if one can establish that $P(k + 1)$ is true from the truth of $P(k)$, then $P(n)$ is true for all n. Intuitively, this means that from the truth of $P(1)$, we can infer that $P(2)$ is true and, from the truth of $P(2)$, infer that $P(3)$ is true, etc. Our proof will be the first example of a proof by contradiction. Many theorems have the form "If A is true, then B is true"; this example is of that type. To prove such a theorem by contradiction, one assumes that A is true and B is false and shows that this leads to a logical contradiction, often showing that A is false. This means that the assumption "A is true and B is false" can't be true, and hence the theorem must be true.

Proof. Assume that $P(n)$ is a statement about $n \in J$ such that

a. $P(1)$ is true, and
b. for each $k \in J$, if $P(k)$ is true, then $P(k + 1)$ is true.

In addition, assume that $P(n)$ is *not* true for some $n \in J$. (This is the "A is true, B is false" part of the proof by contradiction). Since $P(n)$ is false for some $n \in J$, the set

$$S = \{n : P(n) \text{ is false}\}$$

is a nonvoid subset of J. As such, by the well-ordering principle, S has a smallest member, call it n_0. Since $P(1)$ is true, $n_0 > 1$. Since n_0 is the smallest member of S, $P(n_0)$ is false and $P(n_0 - 1)$ is true. But remember, if $P(k)$ is true, then $P(k + 1)$ is true! So, if we let $k = n_0 - 1$, then $P(k)$ is true, but $k + 1 = n_0$ and $P(k + 1)$ is false. This is a contradiction. Hence, by the method of proof by contradiction, the theorem is true.

In the pages to come, you will see more proofs by contradiction, so now is a good time to examine the process by reading the proof several times for maximum understanding.

■ **Example 0.9** The following identity will be needed in Chapter 6:

$$\text{For } n \in J, \quad 1 + x + x^2 + \cdots + x^n = \frac{1 - x^{n+1}}{1 - x} \quad \text{if } x \neq 1.$$

We will use mathematical induction to verify this identity. Let $P(n)$ be the statement:

$$1 + x + x^2 + \cdots + x^n = \frac{1 - x^{n+1}}{1 - x} \quad \text{if } x \neq 1.$$

a. First check to see that $P(1)$ is true. If $n = 1$ the left member of this identity is $1 + x$ and the right member is $\dfrac{1 - x^2}{1 - x} = 1 + x$ if $x \neq 1$. Thus $P(1)$ is true.
b. Now assume that $P(k)$ is true and try to show that this implies that $P(k + 1)$ is true. So we have

$$1 + x + x^2 + \cdots + x^k = \frac{1 - x^{k+1}}{1 - x} \quad \text{if } x \neq 1,$$

so

$$\begin{aligned} 1 + x + x^2 + \cdots + x^k + x^{k+1} &= \frac{1 - x^{k+1}}{1 - x} + x^{k+1} \\ &\qquad \text{(because we assumed } P(k) \text{ is true)} \\ &= \frac{1 - x^{k+1} + x^{k+1}(1 - x)}{1 - x} \\ &= \frac{1 - x^{k+2}}{1 - x} \quad \text{if } x \neq 1 \end{aligned}$$

This last equality asserts that $P(k + 1)$ is true. Thus, from the truth of $P(k)$, we can prove that $P(k + 1)$ is true.

By the principle of mathematical induction, $P(n)$ is true for all $n \in J$. This was our goal. ■

Under some circumstances, we need what is sometimes referred to as the *second principle of mathematical induction.*

0.9 THEOREM (SECOND PRINCIPLE OF MATHEMATICAL INDUCTION) Suppose that $P(n)$ is a statement for each $n \in J$. If

a. $P(1), P(2), \ldots, P(m)$ is true, and
b. for $k > m$, if $P(i)$ is true for $1 \leq i \leq k$, then $P(k + 1)$ is true.

then $P(n)$ is true for all $n \in J$.

Proof. We leave the proof as Exercise 23. The proof is very similar to that of Theorem 0.8 and utilizes the well-ordering principle.

The function defined in Example 0.10 is defined by *recursion*; that is, some initial values of f are given—in this case, $f(1)$ and $f(2)$—and then, for larger n, $f(n)$ is given in terms of the values $f(k)$ for some $k < n$. In this case $f(n) = \dfrac{f(n-1) + f(n-2)}{2}$. We will use the second principle of mathematical induction to prove that $f(n) = 2 + \left(-\frac{1}{2}\right)^{n-1}$ for all $n \in J$.

■ **Example 0.10** Define $f\colon J \to R$ as follows:

$$f(1) = 3, \quad f(2) = \frac{3}{2}, \quad \text{and} \quad \text{for } n \geq 3 \quad f(n) = \frac{f(n-1) + f(n-2)}{2}$$

We want to prove that $f(n) = 2 + \left(-\frac{1}{2}\right)^{n-1}$ for all $n \in J$. We will utilize Theorem 0.9. First of all,

a. We will show that $P(1)$ and $P(2)$ are true. For $n = 1$ the formula gives

$$2 + \left(-\frac{1}{2}\right)^{0} = 2 + 1 = 3 = f(1).$$

For $n = 2$, the formula gives

$$2 + \left(-\frac{1}{2}\right) = 2 - \frac{1}{2} = \frac{3}{2} = f(2).$$

So, for $n = 1$ and $n = 2$, $f(n) = 2 + \left(-\frac{1}{2}\right)^{n-1}$.

b. Assume now that $f(i) = 2 + \left(-\frac{1}{2}\right)^{i-1}$ for $1 \leq i \leq k$. Then

$$f(k+1) = \frac{f(k) + f(k-1)}{2} = \frac{1}{2}\left[2 + \left(-\frac{1}{2}\right)^{k-1} + 2 + \left(-\frac{1}{2}\right)^{k-2}\right]$$
$$= \frac{1}{2}\left[4 + \frac{1}{2}\left(-\frac{1}{2}\right)^{k-2}\right] = 2 + \left(-\frac{1}{2}\right)^{k}.$$

Thus the formula holds for $n = k + 1$.

By the second principle of mathematical induction, $f(n) = 2 + \left(-\frac{1}{2}\right)^{n-1}$ for all $n \in J$. ■

The proofs of both versions of mathematical induction depend on the well-ordering principle of the set of positive integers. Both the well-ordering principle and the two principles of mathematical induction can be modified easily as follows.

WELL-ORDERING PRINCIPLE (Modified) If S is any nonempty subset of Z (remember, Z is the set of all integers) such that S has a smallest member, then any nonempty subset of S has a smallest member.

Proof. Suppose S is a nonempty subset of Z and s_0 is the smallest member of S. Let A be a nonvoid subset of S. If $A \subset J$, then the first version of the well-ordering principle gives the result. If A is not a subset of J, then $s_0 < 1$. Let $T = \{s_0, s_0 + 1, s_0 + 2, \ldots, 0\}$. Then $A \cap T$ is a nonvoid finite set, and the smallest member of A is the smallest member of $A \cap T$.

If you reexamine the proof of Theorem 0.8, you will see that the argument relied in part on the fact that, if $P(n)$ was false for some n, then there was a smallest such n and, since $P(1)$ was true, that particular n was greater than 1. Using the modified well-ordering principle, we can prove the following modification of Theorem 0.8.

0.10 THEOREM (MODIFIED PRINCIPLE OF MATHEMATICAL INDUCTION) Let $P(n)$ be a statement for $n \in Z$. If

a. $P(n_0)$ is true, and
b. for $k \in Z$, $k \geq n_0$, if $P(k)$ is true, then $P(k + 1)$ is true, then

$P(n)$ is true for all $n \geq n_0$.

Proof. The proof is left as Exercise 27.

Exercise 28 asks you to prove a modified version of Theorem 0.9. The next example shows how Theorem 0.10 might be applied.

■ **Example 0.11** Consider the statement $P(n) : 2n + 1 < 2^n$. Try a few values of n:

$$\begin{array}{lll} n = 1 & 2 \cdot 1 + 1 = 3 \nless 2 = 2^1 & P(1) \text{ is false.} \\ n = 2 & 2 \cdot 2 + 1 = 5 \nless 4 = 2^2 & P(2) \text{ is false.} \\ n = 3 & 3 \cdot 2 + 1 = 7 < 8 = 2^3 & \text{Aha! } P(3) \text{ is true.} \end{array}$$

So, we will apply the modified version of mathematical induction with $n_0 = 3$.

a. $P(3)$ is true (see above).
b. Assume $k \geqslant 3$ and $P(k)$ is true. Then

$$\begin{aligned} 2(k + 1) + 1 = 2k + 1 + 2 &< 2^k + 2 \\ &< 2^k + 2^k = 2^{k+1}. \end{aligned}$$

Thus, $P(k + 1)$ is true.

Therefore, $P(n)$ is true for all $n \geqslant 3$. ■

0.4 EQUIVALENT AND COUNTABLE SETS

Consider now the sets $A = \{1, 2, 3, 4, 5\}$ and $B = \{2, 4, 6, 8, 10\}$. Is is clear that $A \neq B$, and in fact $A \not\subset B$ and $B \not\subset A$, so there seems to be no way of comparing the two sets. However, a similarity exists: Both sets have exactly five elements, a result we easily obtain by counting the members of each set. If the sets were quite large, counting elements would be a difficult task; in fact, if the sets were infinite, then we would be out of luck. The purpose of the following discussion is to consider a way of comparing the size of two sets without the necessity of using the natural numbers as a guide for counting. An obvious way of showing that A and B have the same number of elements is to pair each element of A with an element of B and observe that, when this is accomplished, there are no elements of B remaining. The following is one such way of doing this: (1, 2), (2, 4), (3, 6), (4, 8), and (5, 10). The sophisticated reader will recognize that this amounts to defining a 1–1 function from A onto B, in this case the function being easily given by the formula $f(x) = 2x$ for each $x \in A$. There are, of course, many other ways of pairing the elements of A with the elements of B. The reader might be interested in computing how many there are.

DEFINITION If A and B are sets, we say that A is *equivalent* to B, written $A \sim B$, iff there is a 1–1 function f from A onto B.

To the reader who has sampled the flavor of modern mathematics, the word *equivalent* carries certain implications. Before considering some examples relating to this definition, let us prove a theorem that justifies the usage.

0.11 THEOREM Let A, B, and C be sets. Then

1. $A \sim A$
2. If $A \sim B$, then $B \sim A$.
3. If $A \sim B$ and $B \sim C$, then $A \sim C$.

Proof. 1. To show $A \sim A$, we must exhibit a 1–1 function f from A onto A. It seems reasonable to try 1_A. Now if

$$1_A(a_1) = 1_A(a_2),$$

then

$$a_1 = 1_A(a_1) = 1_A(a_2) = a_2;$$

hence 1_A is 1–1. It is clear that im $1_A = A$ since, for any $a \in A$, $1_A(a) = a$. Thus $A \sim A$.

2. Suppose $A \sim B$. Then there is a 1–1 function f from A onto B. To show $B \sim A$, one must find a 1–1 function g from B onto A. The discerning reader should now observe that f^{-1} is the logical candidate. It has already been shown that f^{-1} is 1–1, dom f^{-1} = im $f = B$, and im f^{-1} = dom $f = A$; hence, f^{-1} is a 1–1 function from B onto A. Therefore $B \sim A$.

3. Assume $A \sim B$ and $B \sim C$. There are 1–1 functions f from A onto B and g from B onto C. We seek a 1–1 function from A onto C. The only reasonable way to obtain a function from A onto C is to consider the composition of g by f, namely $g \circ f$. We know that dom $(g \circ f) = A$, and it remains to be proved that $(g \circ f)$ is 1–1 and that im $(g \circ f) = C$. This can be done directly, but it's worthwhile to do it via two lemmas.

0.12 LEMMA Suppose $f: A \to B$ and $g: B \to C$. If f and g are 1–1, then $g \circ f$ is 1–1.

Proof. Suppose f and g are 1–1 and that $a_1, a_2 \in A$ with $(g \circ f)(a_1) = (g \circ f)(a_2)$. Now

$$g(f(a_1)) = (g \circ f)(a_1) = (g \circ f)(a_2) = g(f(a_2)),$$

and, since g is 1–1, $f(a_1) = f(a_2)$. In like fashion, the fact that f is 1–1 guarantees that $a_1 = a_2$. Therefore, $g \circ f$ is 1–1.

0.13 LEMMA Suppose $f: A \to B$ and $g: B \to C$. If im $f = B$ and im $g = C$, then im $(g \circ f) = C$.

Proof. Suppose $c \in C$. Since im $g = C$, there is $b \in B$ such that $g(b) = c$; and, since im $f = B$, there is a $a \in A$ such that $f(a) = b$. Now

$$(g \circ f)(a) = g(f(a)) = g(b) = c.$$

Thus $C \subset \text{im}\,(g \circ f)$. Conversely, if $a \in A$, then $f(a) \in B$ and

$$(g \circ f)(a) = g(f(a)) \in \text{im}\, g = C,$$

so $\text{im}\,(g \circ f) \subset C$. Thus, $\text{im}\,(g \circ f) = C$.

With these two lemmas we can now complete the proof of Theorem 0.11. Since $f\colon A \to B$ and $g\colon B \to C$ are 1–1, then $g \circ f$ is 1–1; and, since f is onto B and g is onto C, then $g \circ f$ is onto C. Therefore $A \sim C$.

There is much more to be said concerning the equivalence of two sets. We shall consider only a few instances that are pertinent to later topics. Let us first consider some examples.

Let J denote the set of positive integers, and let E denote the set of even positive integers. Define $f\colon J \to E$ by $f(n) = 2n$ for each $n \in J$. Now, if $f(n) = f(m)$, then

$$2n = f(n) = f(m) = 2m;$$

hence $n = m$, so f is 1–1. Since each even positive integer may be written in the form $2n$ where $n \in J$, it is clear that f is 1–1 onto E; hence $J \sim E$. This may seem to be a peculiar state of affairs—the set J being equivalent to one of its proper subsets. However, this is typical of such sets; in fact, one may define an infinite set to be any set that is equivalent to one of its proper subsets.

Let J denote the set of positive integers, and let Z denote the set of all integers. Define $f\colon J \to Z$ by the following set of rules: If $n \in J$ and n is even,

$$f(n) = \frac{n}{2};$$

and, if $n \in J$ and n is odd,

$$f(n) = (-1)\frac{n-1}{2}.$$

To show $J \sim Z$, we need to show that f is a 1–1 function from J onto Z. By construction of f, it is clear that $\text{dom}\, f = J$ and $\text{im}\, f \subset Z$. If $m \in Z$ and m is positive, then $2m \in J$ and $f(2m) = m$. If $m \in Z$ and m is zero or negative, then $1 - 2m \in J$ and

$$f(1 - 2m) = (-1)\frac{(1 - 2m) - 1}{2} = m,$$

since $1 - 2m$ is odd. Thus f is onto Z and it remains to show that f is 1–1. For each $n \in J$ with $n \neq 1$, $f(n) > 0$ if n is even and $f(n) < 0$ if n is odd [$f(1) = 0$]. Thus, if $f(s) = f(r) = a$, then s and r must be both 1, both even, or both odd as a is zero, positive, or negative. If s and r are both even, then

$$s = 2\left(\frac{s}{2}\right) = 2f(s) = 2f(r) = 2\left(\frac{r}{2}\right) = r.$$

If s and r are both odd, then

$$s = (-2)\left[-\left(\frac{s-1}{2}\right)\right] + 1 = -2[f(s)] + 1 = -2[f(r)] + 1 = r.$$

Hence f is 1–1.

We have shown that $J \sim E$ and $J \sim Z$, so, by use of Theorem 0.11, $Z \sim E$. Sets equivalent to J are of sufficient importance to deserve a special name.

DEFINITION A set S is *countably infinite* iff S is equivalent to the set J of positive integers. A set S is *finite* iff S is empty or there is $n \in J$ such that S is equivalent to the set $\{1, 2, 3, \ldots, n\}$. A set is *countable* if it is either finite or countably infinite. A set is *infinite* if it is not finite. A set is *uncountable* if it is not countable.

We will see examples of uncountable sets later, but first we consider more examples of countable sets. Three examples of countable sets have been exhibited—namely E, Z, and, of course, J itself since $J \sim J$. You might find it instructive to stop now and discover a few more examples for yourself. As a start, try to prove that the set of odd positive integers is countable.

The following theorem shows that, in a certain sense, countably infinite sets are the smallest infinite sets.

0.14 THEOREM Any infinite subset of J is countably infinite.

Proof. Let S be an infinite subset of J. We will define a function $f: J \to S$ recursively as follows: Since S is a nonempty subset of J, it has a least member; define $f(1)$ to be that least member. After $f(1), f(2), \ldots, f(k)$ have been defined, we define $f(k + 1)$ to be the least member of $S \setminus \{f(1), f(2), \ldots, f(k)\}$. This is always possible since $S \setminus \{f(1), f(2), \ldots, f(k)\}$ is always a nonempty subset of J. The function is certainly 1–1 by its construction. Also, if $n \in S$, then there is $m \in J$ such that n is the smallest element of $S \setminus \{f(1), f(2), \ldots, f(m)\}$, and hence $n = f(m + 1)$. Thus f is a 1–1 function from J onto S. This means that S is countably infinite.

0.15 COROLLARY Any subset of a countable set is countable.

Proof. The proof is left as exercise 30.

A consequence of Corollary 0.15 is worth noting here. Let S be a set and suppose there is a 1–1 function $f: S \to J$. Then S is equivalent to $f(S)$, a subset of J. But $f(S)$ is countable since it is a countable subset of a countable set, hence S is countable. In other words, to show that a set is countable it is sufficient to exhibit a 1–1 function from that set *into* J. That technique is used in the next theorem.

0.16 THEOREM If A and B are countable, then $A \times B$ is countable.

Proof. Suppose A and B are countable. Then there are 1–1 functions $f: A \to J$ and $g: B \to J$. Define $h: A \times B \to J$ by

$$h(a, b) = 2^{f(a)}3^{g(b)}.$$

Since f and g are 1–1, h must be 1–1 by the Unique Factorization Theorem. By the remarks preceding the theorem, $A \times B$ is countable.

Recall that *countable* means countably infinite or finite. You may have noted that the fact that S is countable iff it is equivalent to a subset of J allows us to avoid the difficulty of treating the finite and infinite cases separately.

■ **Example 0.12** Let S be the set of all rational numbers. Now each rational number can be written uniquely as $\frac{a}{b}$, where a and b are integers, $b > 0$, a and b are relatively prime (that is, a and b have no common divisors greater than 1), and, if $a = 0$, $b = 1$. We define a function $f: S \to J \times J$ as follows:

$$f(r) = f\left(\frac{a}{b}\right) = (a, b) \text{ where } \frac{a}{b} \text{ is the unique representation}$$

of the rational number r. This function is 1–1 from S onto a subset of $J \times J$, and, by Theorem 0.16, $J \times J$ is countable; hence, any subset of $J \times J$ is countable by Corollary 0.15. But, since S is equivalent to a subset of $J \times J$, then S is countable. ■

There are many ways of combining countable sets to obtain countable sets. Theorem 0.16 gives one such way, and the next theorem gives another method. This theorem will be useful in Chapter 2.

0.17 THEOREM Let S be a nonempty subset of J. Let $\{A_s\}_{s \in S}$ be an indexed family of countable sets. Then $\bigcup_{s \in S} A_s$ is a countable set.

Proof. Since each A_s is countable, for each $s \in S$, there is a 1–1 function $f_s: A_s \to J$. If $x \in \bigcup_{s \in S} A_s$, then there is a smallest member m of S such that $x \in A_m$. We will attempt to define a function $f: \bigcup_{s \in S} A_s \to J \times J$ as follows: $f(x) = (m, f_m(x))$, where m is the least integer in S such that $x \in A_m$. Since $x \in \bigcup_{s \in S} A_s$, it must belong to A_m for some $m \in S$; hence there is a smallest such m by the well-ordering principle. Thus f is a well-defined function—that is, the rule for finding $f(x)$ is unambiguous. We now need to show that f is 1–1. Suppose

$$f(x) = f(y) = (m, n).$$

Thus $x \in A_m$ and $y \in A_m$, hence $n = f_m(x) = f_m(y)$. Because f_m is 1–1, $x = y$. Thus f is a 1–1 function. We have constructed a 1–1 function f from $\bigcup_{s \in S} A_s$ onto a subset of $J \times J$, hence $\bigcup_{s \in S} A_s$ is a countable set.

Our preoccupation with countable sets would seem bizarre if all infinite sets were countable. However, there are lots of uncountable sets. The set of real numbers is a familiar example, although you may not be familiar with the fact that R is uncountable. In Exercise 37 you are asked to prove that, for any set A, $P(A)$—the set of all subsets of A—is *not* equivalent to A. Hence, the set of all subsets of J is another example of an uncountable set.

There are several ways to prove that R is uncountable; we will wait until Chapter 3 to give a proof. However, we can show an example of a set equivalent to R.

■ **Example 0.13** We will show that R is equivalent to $(0, 1)$. The easiest way to accomplish this is to use a function familiar to anyone who has studied trigonometry. Define $f\colon (0, 1) \to R$ by $f(x) = \tan\left(\pi x - \dfrac{\pi}{2}\right)$. This gives a 1–1 function from $(0, 1)$ to R. ■

See the project at the end of this chapter for other sets equivalent to $(0, 1)$.

0.5 REAL NUMBERS

You probably have a good knowledge of the set of rational numbers, the order relation on that set, and the operations of addition and multiplication. However, the set of rational numbers is not sufficient to support all of mathematics. The following theorem shows that an equation as simple as $x^2 = 2$ has no solution in the set of rational numbers.

0.18 THEOREM There is no rational number whose square is 2.

Proof. Suppose there are positive integers p and q such that $2 = (p/q)^2$. Assume further that p and q are relatively prime—that is, their greatest common divisor is 1. Thus we have $2q^2 = p^2$, so p is even (since p odd implies p^2 odd). We may write $p = 2r$ where r is an integer. Then $2q^2 = p^2 = 4r^2$, so $q^2 = 2r^2$, and hence q is even. Thus, both p and q are even, contrary to the assumption that p and q are relatively prime.

Rather than construct the set of real numbers, we shall assume the existence of that set and postulate the properties that we will need. For those who wish a careful and rigorous development of the real numbers, see Edmund Landau's *Foundations of Analysis, 2nd Ed.* (New York: Chelsea Publishing Co., 1960).

We assume the existence of a set R, called the set of *real numbers,* that satisfies the following axioms. There are functions $+\colon R \times R \to R$ and $\cdot\colon R \times R \to R$ and a relation $<$ on R such that, for all $x, y, z \in R$, we have

1. $(x + y) + z = x + (y + z)$; $(x \cdot y) \cdot z = x \cdot (y \cdot z)$.
2. $x + y = y + x$; $x \cdot y = y \cdot x$.
3. $x \cdot (y + z) = (x \cdot y) + (x \cdot z)$.
4. There is a unique element $0 \in R$ such that $0 + x = x$ for all $x \in R$.
5. For each $x \in R$, there is a unique $y \in R$ such that $x + y = 0$, and we write $y = -x$.
6. There is a unique element $1 \in R$ such that $x \cdot 1 = x$ for all $x \in R$.
7. For each $x \in R$ with $x \neq 0$, there is a unique element $y \in R$ such that $x \cdot y = 1$, and we write $y = x^{-1}$ or $y = \frac{1}{x}$.
8. $x < y$ implies $x + z < y + z$.
9. $x < y$ and $y < z$ implies $x < z$.
10. For $x, y \in R$, exactly one of the following is true: $x < y$, $y < x$, or $x = y$.
11. $x < y$ and $z > 0$ implies $xz < yz$.

This list of properties assures us that the system $(R, +, \cdot, <)$ is an *ordered field*. But we need more.

A set $S \subset R$ is said to be *bounded from above* (*below*) if there is a real number M such that, for all $x \in S$, $x \leqslant M$ $(x \geqslant M)$. A set is *bounded* if it is bounded both from above and from below. If $S \subset R$, a real number M is an *upper bound* (*lower bound*) for S if for all $x \in S$, $x \leqslant M$ $(x \geqslant M)$. If a nonvoid set S is finite, then S is obviously bounded; and, in fact, among the members of S there must be a largest, which we denote by *max S*, and a smallest, which we denote by *min S*. For an infinite set S, it is not necessarily true that S has either a largest or a smallest number.

Let S be a set of real numbers bounded from above. A real number a is a *least upper bound* for S if a is an upper bound for S having the property that, if b is also an upper bound for S, then $a \leqslant b$. If S is bounded from below, then a real number a is a *greatest lower bound* for S if a is a lower bound for S having the property that, if b is any lower bound for S, then $b \leqslant a$. It is immediately clear that if S has a least upper bound a, then it is unique; we write $a = \text{l.u.b. } S$ or $a = \sup S$. The abbreviation *sup* is from the word *supremum*, often used as a synonym for least upper bound. Likewise, if S has a greatest lower bound c, then it is unique; we write $c = \text{g.l.b. } S$ or $c = \inf S$. *Inf* come from *infimum*, a synonym for greatest lower bound.

The next property of the set of real numbers is very important to the analyst. It is called the *least upper bound property*.

12. Every nonempty set of real numbers that is bounded from above has a least upper bound.

Axioms 1–12 tell us that $(R, +, \cdot, <)$ is a *complete ordered field*. Axioms 1–7 are the axioms for a field, axioms 8–11 indicate that the order relation interacts with addition and multiplication in the proper manner, and axiom 12 is required that the ordered field be complete. The set of rational numbers with $+$, $\cdot$, and $<$ is an ordered field, but *not* a complete ordered field. The set $\{x : x \text{ is a rational number and } x^2 < 2\}$ has no least upper bound in the set of rational numbers. The proof of this statement is left as Exercise 43. However, before you tackle that exercise, you will need some more operating machinery. Axioms 1–12 were designed to be a minimum set of axioms from which we can prove other relations that will be useful later. The next theorem offers some of those relations.

0.19 THEOREM Let x, y, and z be any real numbers. Then

i. If $x < y$, then $-y < -x$.
ii. $0 < 1$
iii. If $0 < x < y$, then $0 < \frac{1}{y} < \frac{1}{x}$.
iv. If $x < y$ and $z < 0$, then $yz < xz$.

Proof. i. Assume $x < y$. Then $x + (-x - y) < y + (-x - y)$ by axiom 7. Removing parentheses and simplifying yields $-y < -x$.

ii. This statement may seem out of place and perhaps ridiculous, but the result is needed in the proof of (iii). Besides, this statement should be a result of the axioms. By axiom 10, exactly one of the following holds: $0 = 1$, $0 < 1$, or $1 < 0$. We know that $0 \neq 1$, so consider the possibility $1 < 0$. Then by (i) above, $0 < -1$. But then, by axiom 11, $0 \cdot (-1) < (-1)(-1)$, or $0 < 1$. This contradicts the assumption $1 < 0$. The only remaining case is $0 < 1$. Since one of the three statements—$0 = 1$, $0 < 1$, or $1 < 0$—must be true, it is the case that $0 < 1$.

iii. Suppose $0 < x < y$. Now, by axiom 10, $\frac{1}{x} = 0$, $\frac{1}{x} < 0$, or $\frac{1}{x} > 0$. Since $x \cdot \frac{1}{x} = 1$, we rule out $\frac{1}{x} = 0$. Suppose that $\frac{1}{x} < 0$. Then, since $x > 0$, $1 = x \cdot \frac{1}{x} < x \cdot 0 = 0$, which contradicts (ii). Therefore $\frac{1}{x}$ and $\frac{1}{y}$ are both greater than zero. But then $0 < x\left(\frac{1}{x}\right)\left(\frac{1}{y}\right) < y\left(\frac{1}{x}\right)\left(\frac{1}{y}\right)$, or $0 < \frac{1}{y} < \frac{1}{x}$.

iv. Suppose $x < y$ and $z < 0$. Then $0 < -z$, and by axiom 11 $-zx < -zy$ and then by (i) $zy < zx$.

Axiom 12 assures the existence of a least upper bound for any set bounded from above. As we will have need of greatest lower bounds also, the next theorem is important, but easy to prove from axiom 12 and the result of Theorem 0.19.

0.20 THEOREM If S is a nonempty set of real numbers that is bounded from below, then S has a greatest lower bound.

Proof. Let $T = \{t : t = -s \text{ for some } s \in S\}$. If M is any lower bound for S, then $M \leqslant s$ for all $s \in S$; hence, $-s \leqslant -M$ for all $s \in S$, so $-M$ is an upper bound for T. Similarly, if K is an upper bound for T, then $-K$ is a lower bound for S. Since S is bounded from below, T is bounded from above, and by axiom 12, T has a least upper bound, call it B. We claim that $-B$ is the greatest lower bound for S. Certainly $-B$ is a lower bound for S. Let C be any other lower bound for S. Then $-C$ is an upper bound for T, hence $B \leqslant -C$ since B is the least upper bound for T. But this implies that $C \leqslant -B$. Thus $-B$ is the greatest lower bound for S.

We will need to know a bit about how the integers and rational numbers are situated in the system of real numbers. The next few theorems are intended to give us some of that knowledge. Some of the results may seem obvious, but it is important to see that they follow from the axioms.

0.21 THEOREM Let x be any real number. Then there is an integer n such that $n \leqslant x < n + 1$.

Proof. Let $A = \{n : n \in Z \text{ and } n \leqslant x\}$. We will consider two cases—A empty and A nonempty. Assume first that A is nonempty. Since A is bounded from above and is nonempty, A has a least upper bound, call it n_0. Therefore $n_0 - 1$ is not an upper bound for A, hence there is $m \in A$ such that $n_0 - 1 < m \leqslant n_0$. But this implies that $n_0 < m + 1$. Since n_0 is an upper bound for A, then $m + 1 \notin A$; i.e., $x < m + 1$. Therefore, we have found an integer m such that $m \leqslant x < m + 1$.

We now consider the case in which A is empty. Then the set $B = \{n : n > x \text{ and } n \in Z\}$ is not empty. B is bounded from below and nonempty, hence B has a greatest lower bound, call it b_0. Arguing as above, there is an integer $m \in B$ such that $b_0 \leqslant m < b_0 + 1$. Then $m - 1 < b_0$ and so $m - 1 \notin B$. Thus we have found an integer m such that $m - 1 \leqslant x < m$. The integer $m - 1$ meets our requirement. By the way, of course, A can't be empty, but it had to be proven.

We have stated a list of axioms that gives the set of real numbers as a complete ordered field. That's fine and we have been successful in verifying some familiar identities from those axioms. The model that we have for the set of real numbers is the number line or, as it is more commonly called, the *real* line. In an early algebra course, you learned all about the correspondence between rational numbers and points on the line, although the concept of a real number may have seemed a bit vague at the time. We hope the uncertainty will fade as our development continues. However, do not lose sight of the real line as a model for the set of real numbers. Thus, $x < y$ means that x is located to the left of y on the real line. Theorem 0.21 tells us that every real number is located between consecutive integers n and $n + 1$.

In Section 0.1, we identified some special sets.

1. $[a, b] = \{x : x \in R \text{ and } a \leqslant x \leqslant b\}$
2. $[a, b) = \{x : x \in R \text{ and } a \leqslant x < b\}$

3. $(a, b] = \{x : x \in R \text{ and } a < x \leqslant b\}$
4. $(a, b) = \{x : x \in R \text{ and } a < x < b\}$

The set $[a, b]$ is referred to as a *closed* interval, and the set (a, b) is referred to as an *open* interval. The reasons for these choices of adjectives will be apparent later.

The next theorem states that every open interval (a, b) with $a < b$ contains a rational number. In more technical terms, it states that the set of rational numbers is *dense* in R.

0.22 THEOREM Between any two real numbers, there is a rational number.

Proof. Let x and y be real numbers with $x < y$. By Theorem 0.21, there is an integer N such that $0 < \dfrac{1}{y - x} < N$. Note in particular that $1 < (y - x)N$. Again by Theorem 0.21, there is an integer n such that $n \leqslant Nx < n + 1$. Then $n + 1 \leqslant Nx + 1 < Nx + N(y - x) = Ny$. So we have

$$Nx < n + 1 < Ny$$

and since $N > 0$,

$$x < \frac{n + 1}{N} < y.$$

The rational number $\dfrac{n + 1}{N}$ is the required number.

It is worth noting that Theorem 0.21 implies something that sounds to be stronger. If x and y are real numbers and there are only a finite number of rational numbers between x and y—let $r_1 < r_2 < r_2 < \cdots < r_m$ be the list—then there are *no* rational numbers between r_1 and r_2, for example. Thus, between any two real numbers, there are infinitely many rational numbers.

We indicated in Section 0.4 that the set of real numbers is uncountable, whereas the set of rational numbers is countable. This means that there are lots of real numbers that are not rational numbers. A real number that is not a rational number is called an *irrational* number. Theorem 0.18 asserts that $\sqrt{2}$ is such a number. As we will see soon, the set of irrational numbers is also dense in R; that is, every nonempty open interval contains an irrational number. As argued above, this means that every nonempty open interval contains infinitely many irrational numbers. To prove all of this, we need a ready supply of irrational numbers. See the next theorem.

0.23 THEOREM If $p > 0$ is an integer but not the square of an integer, then $\sqrt{p}$ is an irrational number.

Proof. This proof will use the well-ordering principle and contradiction. Assume that p is not the square of an integer and that $\sqrt{p}$ is a rational number. Thus $\sqrt{p}$ is not an integer, but $n\sqrt{p}$ is an integer for some integer n. Let $A = \{n : n \in J \text{ and } n\sqrt{p} \text{ is an integer}\}$. A is nonempty, hence A has a smallest member, call it m. There is $k \in J$ such that $0 < k < \sqrt{p} < k + 1$. Note that, since $p > 0$ and p is not the square of an integer, then $p > 1$; hence $\sqrt{p} > 1$. Now consider the number

$$x = m\sqrt{p} - km.$$

We need to note several things about x. First, $0 < k < \sqrt{p} < k + 1$ implies that

$$\text{a. } 0 < \sqrt{p} - k < 1,$$

and hence

$$\text{b. } 0 < x = m(\sqrt{p} - k) < m.$$

Moreover,

$$\text{c. } x\sqrt{p} = mp - mk\sqrt{p}.$$

Both x and $x\sqrt{p}$ are integers and $x > 0$, hence $x \in A$ contrary to m being the smallest member of A. Thus, by contradiction, $\sqrt{p}$ is not a rational number.

Theorem 0.23 guarantees the existence of lots of irrational numbers, but don't be fooled. Not all irrational numbers are found by square roots, cube roots, etc. See Exercise 36 for the definition of an *algebraic number.* You are asked to prove that the set of algebraic numbers is countable. It includes all the roots referred to above. There are then many real numbers that are *not* algebraic; such numbers are called *transcendental.* The numbers π and e, the base for the natural logarithms, are examples of transcendental numbers.

The next theorem we will prove states that the set of irrational numbers is dense in R. In other words, between any two real numbers there is an irrational number. You might want to look at the strategy of the proof of Theorem 0.22 before proceeding. The proof of the next theorem is very similar to that of Theorem 0.22.

0.24 THEOREM Between any two real numbers, there is an irrational number.

Proof. Let x and y be any real numbers with $x < y$. In the proof of Theorem 0.22, we needed an integer N such that $\dfrac{1}{y - x} < N$. For this proof, we need a nonsquare M such that $\dfrac{1}{(y - x)^2} < M$, so a variation on Theorem 0.21 is needed. Given a real number z, we need an integer that is not a square and that is larger than z. The next lemma does that job for us.

LEMMA If z is any real number, there is a positive integer P such that $z < P$ and P is not the square of an integer.

Proof. If $z < 0$, let $M = 1$, otherwise by Theorem 0.21, there is a positive integer M such that $z < M$. If M is a square, then $M + 1$ is not a square. Can you prove this? Try proving it by contradiction. If $M = n^2$ and $M + 1 = m^2$, then $1 = M + 1 - M = m^2 - n^2 = (m + n)(m - n)$. Thus either M or $M + 1$ meets the needs of the lemmas.

Now we may find an integer M such that M is not a square and $0 < \dfrac{1}{(y - x)^2} < M$. Therefore, $0 < \dfrac{1}{\sqrt{M}} < y - x$. By Theorem 0.21, there is an integer n such that $n \leq \sqrt{M}x < n + 1$. Then

$$n + 1 \leq \sqrt{M}x + 1 < \sqrt{M}x + \sqrt{M}(y - x) = \sqrt{M}y.$$

Therefore,

$$x < \frac{n + 1}{\sqrt{M}} < y,$$

and $\dfrac{n + 1}{\sqrt{M}}$ is the required irrational number since $\sqrt{M}$ is irrational by Theorem 0.23.

In the study of analysis, the notion of the distance between two points on the real line—that is, the distance between two real numbers—is of prime importance. The best way to describe this distance is in terms of *absolute value*.

If x is any real number, define

$$|x| = \begin{cases} x \text{ if } x \geq 0 \\ -x \text{ if } x < 0. \end{cases}$$

we call $|x|$ the absolute value of x. Note that $|x| = \sqrt{x^2}$ and $x \leq |x|$ for all $x \in R$.

0.25 THEOREM Let a and b be any real numbers. Then

1. $|a + b| \leq |a| + |b|$.
2. $|ab| = |a| \cdot |b|$.
3. $||a| - |b|| \leq |a - b|$.
4. If $\epsilon > 0$, then $|a| < \epsilon$ iff $-\epsilon < a < \epsilon$.

Proof. Perhaps the easiest way to prove this theorem is to use the fact that $|x| = \sqrt{x^2}$.

1. Let a and b be any real numbers. Then

$$0 \leq (a + b)^2 = a^2 + 2ab + b^2 \leq |a|^2 + 2|a||b| + |b|^2 = (|a| + |b|)^2.$$

Therefore

$$|a + b| = \sqrt{(a + b)^2} \leq \sqrt{(|a| + |b|)^2} = |a| + |b|.$$

2. Let a and b be any two real numbers. Then

$$|ab| = \sqrt{(ab)^2} = \sqrt{a^2b^2} = \sqrt{a^2}\sqrt{b^2} = |a||b|.$$

3. By (1) above,

$$|a| = |a - b + b| \leq |a - b| + |b|;$$

hence

$$|a| - |b| \leq |a - b|.$$

In similar fashion, $|b| \leq |a - b| + |a|$; hence $|b| - |a| \leq |a - b|$. Since $||a| - |b||$ is equal to $|b| - |a|$ or $|a| - |b|$, depending on the relative magnitudes of $|a|$ and $|b|$, we have shown that $||a| - |b|| \leq |a - b|$.

4. Suppose $-\epsilon < a < \epsilon$, $\epsilon > 0$. If $a > 0$, then $|a| = a < \epsilon$. If $a \leq 0$, then $|a| = -a$ and $-\epsilon < a$; hence $|a| = -a < \epsilon$. Thus, $-\epsilon < a < \epsilon$, $\epsilon > 0$, implies that $|a| < \epsilon$. If $|a| < \epsilon$ and $a > 0$, then $-\epsilon < 0 < a = |a| < \epsilon$. If $a \leq 0$, then $|a| = -a < \epsilon$, hence $-\epsilon < a \leq 0 < \epsilon$. Thus, $|a| < \epsilon$ implies that $-\epsilon < a < \epsilon$.

It is to be noted that (4) implies that the following statements are equivalent.

1. $|a - b| < \epsilon$.
2. $-\epsilon < a - b < \epsilon$.
3. $b - \epsilon < a < b + \epsilon$.
4. $a - \epsilon < b < a + \epsilon$.

Again, using the real line as the model for the set of real numbers, $|a - b|$ is just the undirected distance from a to b and, for $\epsilon > 0$,

$$(a - \epsilon, a + \epsilon) = \{x : a - \epsilon < x < a + \epsilon\}$$

is the set of points that are within a distance ϵ of a. Sets of this type play a major role in the ensuing chapters.

EXERCISES

0.1 SETS

1. List the elements of each of the following sets:
a. $J \cap [0, 6)$ b. $Z \cap (-6, 2]$
c. $\{1, 2, 3, 4\} \cup \{2, 3, 4, 5\}$
d. $\{1, 2, 3, 4\} \cap \{2, 3, 4, 5\}$

2. Write each of the following in interval notation:
a. $(0, 2) \cap \left(\frac{1}{2}, 1\right)$ b. $[-1, 5] \cup [2, 7]$

3. Prove (vi) of Theorem 0.2.

4. Prove (ii) of Theorem 0.3.

5. Prove that, for all sets A, B, and C, $A \cap B \subset A \subset A \cup C$.

6. If $A \subset B$, prove that $(C \setminus B) \subset (C \setminus A)$. Either prove the converse is true or give a counterexample.

7. Under what conditions does $A \setminus (A \setminus B) = B$?

8. Show that $(A \setminus B) \cup (B \setminus A) = (A \cup B) \setminus (A \cap B)$.

9. Look up Russell's paradox and write a brief summary discussing how it relates to Section 0.1.

10. Describe each of the following sets as the empty set, as R, or in interval notation, as appropriate:
a. $\bigcap_{n=1}^{\infty} \left(-\frac{1}{n}, \frac{1}{n}\right)$ b. $\bigcup_{n=1}^{\infty} (-n, n)$
c. $\bigcap_{n=1}^{\infty} \left(-\frac{1}{n}, 1 + \frac{1}{n}\right)$ d. $\bigcup_{n=1}^{\infty} \left(-\frac{1}{n}, 2 + \frac{1}{n}\right)$

11. Prove (ii) of Theorem 0.4.

12. Use De Morgan's Laws to give a different and simpler description of the following sets:
a. $R \setminus \bigcap_{n=1}^{\infty} \left(-\frac{1}{n}, \frac{1}{n}\right)$ b. $\bigcup_{n=1}^{\infty} \left(R \setminus \left[\frac{1}{n}, 2 + \frac{1}{n}\right]\right)$

0.2 RELATIONS AND FUNCTIONS

13. Define $f: J \to J$ by $f(n) = 2n - 1$ for each $n \in J$. What is im f? Is f 1–1? Is f onto? If f has an inverse, find the domain of the inverse and give a formula for $f^{-1}(n)$.

14. Often, if the domain is not specified, it is assumed to be the set of all real numbers for which the formula for $f(x)$ defines a real number. What is the domain of the function defined by the formula $f(x) = \dfrac{x}{x + 2}$? What is im f? Is f injective? If so, find the inverse.

For Exercises 15–17, let $A = \{1, 2, 3, 4, 5\}$, $B = \{2, 3, 4, 5, 6, 7\}$, *and* $C = \{a, b, c, d, e\}$.

15. Give an example of $f: A \to B$ that is not 1–1.

16. Give an example of $f: A \to B$ that has an inverse, and show the inverse.

17. Give an example of $f: A \to B$, $g: B \to C$ such that $g \circ f$ is 1–1 but g is not 1–1.

***18.** If $f: A \to B$ is 1–1 and $\operatorname{im} f = B$, prove that $(f^{-1} \circ f)(a) = a$ for all $a \in A$ and $(f \circ f^{-1})(b) = b$ for each $b \in B$.

0.3 MATHEMATICAL INDUCTION AND RECURSION

19. Prove that for all $n \in J$, $1 + 2 + \cdots + n = \dfrac{n(n + 1)}{2}$

20. Prove that for all $n \in J$, $1 + 3 + 5 + \cdots + (2n - 1) = n^2$.

21. Prove that $n^3 + 5n$ is divisible by 6 for each $n \in J$.

22. Prove that $n^2 < 2^n$ for $n \in J$, $n \geqslant 5$. (See Example 0.11)

23. Prove the second principle of mathematical induction (Theorem 0.9).

24. Define $f: J \to J$ by $f(1) = 1, f(2) = 2, f(3) = 3$, and
$f(n) = f(n - 1) + f(n - 2) + f(n - 3)$ for $n \geqslant 4$.
Prove that $f(n) \leqslant 2^n$ for all $n \in J$.

25. Define $f: J \to J$ by $f(1) = 2$ and, for $n \geqslant 2$, $f(n) = \sqrt{3 + f(n - 1)}$. Prove that $f(n) < 2.4$ for all $n \in J$. You may want to use your calculator on this exercise.

26. Define $f: J \to J$ by $f(1) = 2, f(2) = -8$, and, for $n \geqslant 3$,
$f(n) = 8f(n - 1) - 15f(n - 2) + 6 \cdot 2^n$.
Prove that, for all $n \in J$, $f(n) = -5 \cdot 3^n + 5^{n-1} + 2^{n+3}$.

27. Prove Theorem 0.10.

***28.** Prove the following modified version of the second principle of mathematical induction: Let $P(n)$ be a statement for each $n \in Z$. If
a. $P(n_0), P(n_0 + 1), \ldots, P(m)$ is true, and
b. for $k \geqslant m$, if $P(i)$ is true for $n_0 \leqslant i \leqslant k$, then $P(k + 1)$ is true,
then $P(n)$ is true for all $n \geq n_0$, $n \in Z$.

29. Define $f(n)$ as follows for $n \in Z$, $n \geqslant 0$. $f(0) = 7$, $f(1) = 4$, and, for $n \geqslant 2$, $f(n) = 6f(n - 2) - f(n - 1)$. Prove that $f(n) = 5 \cdot 2^n + 2(-3)^n$ for all $n \in Z$, $n \geqslant 0$.

0.4 EQUIVALENT AND COUNTABLE SETS

30. Prove Corollary 0.15.

31. Find a 1–1 function f from J onto S where S is the set of all odd integers.

32. Let P_n be the set of all polynomials of degree n with integer coefficients. Prove that P_n is countable. (*Hint:* A proof by induction is one method of approach.)

33. Use Exercise 32 to show that the set of all polynomials with integer coefficients is a countable set.

34. Prove the following generalization of Theorem 0.17: If S is a countable set and $\{A_s\}_{s \in S}$ is an indexed family of countable sets, then $\bigcup_{s \in S} A_s$ is a countable set.

35. For each $p \in P_n$, define $B(p) = \{x : p(x) = 0\}$. Prove that $\bigcup_{p \in P_n} B(p)$ is countable.

36. An *algebraic number* is any number that is a root of a polynomial equation $p(x) = 0$ where the coefficients of p are integers. Show that the set of algebraic numbers is a countable set.

37. For a set A, let $P(A)$ be the set of all subsets of A. Prove that A is *not* equivalent to $P(A)$. [*Hint:* Suppose $f: A \to P(A)$ and define $C = \{x : x \in A \text{ and } x \not\in f(x)\}$. Show $C \not\in \operatorname{im} f$.]

38. Let a, b, c, and d be any real numbers such that $a < b$ and $c < d$. Prove that $[a, b]$ is equivalent to $[c, d]$. (*Hint:* Show that $[a, b]$ is equivalent to $[0, 1]$ first.)

0.5 REAL NUMBERS

***39.** If $x < y$, prove that $x < \frac{x + y}{2} < y$.

***40.** If $x \geq 0$ and $y \geq 0$, prove that $\sqrt{xy} \leq \frac{x + y}{2}$. [*Hint:* Use the fact that $(\sqrt{x} - \sqrt{y})^2 \geq 0$.]

***41.** If $0 < a < b$, prove that $0 < a^2 < b^2$ and $0 < \sqrt{a} < \sqrt{b}$.

42. If x, y, a, and b are greater than zero and $\frac{x}{y} < \frac{a}{b}$, prove that $\frac{x}{y} < \frac{x + a}{y + b} < \frac{a}{b}$.

43. Let $A = \{r : r \text{ is a rational number and } r^2 < 2\}$. Prove that A has no largest member. [*Hint:* If $r^2 < 2$, and $r > 0$, choose a rational number δ such that $0 < \delta < 1$ and $\delta < \frac{2 - r^2}{2r + 1}$. Show that $(r + \delta)^2 < 2$.]

***44.** If $x = \sup A$, show that, for each $\epsilon > 0$, there is $a \in A$ such that $x - \epsilon < a \leq x$.

***45.** If $y = \inf A$, show that, for each $\epsilon > 0$, there is $a \in A$ such that $y \leq a < y + \epsilon$.

PROJECT

The purpose of this project is to show that the open interval $(0, 1)$ is equivalent to the closed interval $[0, 1]$. In the process we will discover that both intervals are equivalent to $[0, 1)$ and $(0, 1]$. It is then easy to generalize to any interval $[a, b]$ with $a < b$.

Define f: $(0, 1) \to R$ as follows:

$$\text{For } n \in J,\ n \geq 2,\ f\left(\frac{1}{n}\right) = \frac{1}{n - 1}$$

and for all other $x \in (0, 1)$, $f(x) = x$.

1. Prove that f is a 1–1 function from $(0, 1)$ into $(0, 1]$.
2. Prove that f is a function from $(0, 1)$ onto $(0, 1]$.
3. Find a 1–1 function from $[0, 1)$ onto $[0, 1]$. You might use the function in #2 with some modifications.
4. Prove that $[0, 1)$ is equivalent to $(0, 1]$.
5. Prove that $(0, 1)$ is equivalent to $[0, 1]$.

CHAPTER 1

Sequences

The aspiring analyst should begin by investigating the folklore of sequences in detail. A vigorous attempt is made in this chapter to remove the cloud of mystery surrounding the inner workings of sequences. A thorough understanding of sequences is invaluable in understanding the rest of this book.

1.1 SEQUENCES AND CONVERGENCE

DEFINITION A *sequence* is a function whose domain is the set of positive integers.

If a is a sequence, it is customary to write $a(n) = a_n$ for each positive integer n and write $a = \{a_n\}_{n=1}^{\infty}$. We call a_n the nth term of the sequence. If $m \neq n$, then a_m and a_n are considered to be different terms of the sequence even when $a_n = a_m$. In this fashion, it is sometimes convenient to describe a sequence by giving a formula for the nth term. For example, the sequence whose nth term is $\frac{1}{n}$ may be written $\{\frac{1}{n}\}_{n=1}^{\infty}$; the fifth term in this sequence is $\frac{1}{5}$; the ninth term is $\frac{1}{9}$, etc. If p_n denotes the nth prime, the sequence $\{p_n\}_{n=1}^{\infty}$ is well defined, although it may be rather time-consuming to determine the 97th term. As a matter of policy, we shall try to use notation that conveys the intended meaning most clearly. The sequences mentioned earlier may also be described as follows:

$$\left\{\frac{1}{n}\right\}_{n=1}^{\infty} = \left\{1, \frac{1}{2}, \frac{1}{3}, \frac{1}{4}, \frac{1}{5}, \ldots\right\} \quad \text{and} \quad \{p_n\}_{n=1}^{\infty} = \{2, 3, 5, 7, 11, 13, \ldots\}.$$

This device for describing a sequence seems more useful, for example, in describing the sequence

$$\left\{\frac{1 + (-1)^n}{2}\right\}_{n=1}^{\infty} = \{0, 1, 0, 1, 0, 1, \ldots\}.$$

Until further notice, all sequences under consideration will be sequences of real numbers. In other words, the range will be a subset of the set of real numbers. This will enable us to dispense with the necessity of referring to such sequences in a special way and still afford us the luxury of using the notion of a sequence in other ways, as is done in Chapter 5.

Before attempting to define the notion of a convergent sequence, let us consider some of the examples we have at hand. Considering the sequence $\{\frac{1}{n}\}_{n=1}^{\infty}$, we observe that by "going out far enough" in the sequence—that is, choosing n sufficiently large—we find terms that are very close to zero. Likewise, in the sequence $\{0, 1, 0, 1, 0, 1, \ldots\}$ we find terms that are very close to zero; in fact many of them are zero. However, in the latter case we discover that however "far out" in the sequence we look—regardless of how large n is chosen—there are terms farther out in the sequence that are not close to zero. For example, if we choose $n = 1{,}000{,}001$, then $a_{1,000,001} = 0$, but $a_{1,000,002} = 1$, which is not very close to zero. This defect is not present in the first example. If we decide on a measure of closeness, say within .025 of zero, it is clear that all terms of the sequence $\{\frac{1}{n}\}_{n=1}^{\infty}$ beyond the 40th term satisfy this criterion. We shall use this idea as a skeleton upon which to build our definition of convergence. In language less precise than we require, a sequence should converge to a real number A iff, when we settle on how close to A we wish to be, all terms of the sequence from some term on are at least that close to A. Let us now state the formal definition of convergence.

DEFINITION A sequence $\{a_n\}_{n=1}^{\infty}$ *converges to a real number* A iff for each $\epsilon > 0$ there is a positive integer N such that for all $n \geqslant N$ we have $|a_n - A| < \epsilon$.

It is worthwhile to point out that the choice of N may depend upon the choice of ϵ. Let us return to the sequence $\{\frac{1}{n}\}_{n=1}^{\infty}$. Intuition leads us to believe that this sequence should converge to zero (we shall prove it a bit later). If this conclusion is correct, then for $\epsilon > 0$, there is N such that, for $n \geqslant N$, $|a_n - 0| = |\frac{1}{n} - 0| = \frac{1}{n} < \epsilon$. For example, if $\epsilon = .025$, then, for $n \geqslant 41$, $|a_n - 0| = \frac{1}{n} \leqslant \frac{1}{41} < .025$. Thus, for $\epsilon = .025$, $N = 41$ satisfied the conditions of the definition. For $\epsilon = .00025$, one may take $N = 4001$. Note that, if N is such that for $n \geqslant N$, $|a_n - A| < \epsilon$ and $N < M$, then, for $n \geqslant M$, $|a_n - A| < \epsilon$. Hence if $\{a_n\}_{n=1}^{\infty}$ converges to A, then for each $\epsilon > 0$, there are many, in fact infinitely many, positive integers N such that for $n \geqslant N$, $|a_n - A| < \epsilon$. However, to prove that $\{a_n\}_{n=1}^{\infty}$ converges to A, it suffices to show that for each $\epsilon > 0$, there is at least one number N with the desired property.

In order to digest this new idea completely, we shall look for other ways of expressing the notion. Recall that, if a and b are real numbers and $\epsilon > 0$, then $|a - b| < \epsilon$ iff $b - \epsilon < a < b + \epsilon$. Thus $\{a_n\}_{n=1}^{\infty}$ converges to A iff for each $\epsilon > 0$ there is a positive integer N such that for $n \geqslant N$ we have $A - \epsilon < a_n < A + \epsilon$. This means that, given $\epsilon > 0$, $A - \epsilon < a_n < A + \epsilon$ is true except for a finite set of subscripts—such as, in the case mentioned above, some subscripts from the set $\{1, \ldots, N - 1\}$.

To facilitate our study of analysis, we must seek more knowledge of the *topology* of the set of real numbers. Knowledge will come gradually as concepts and facts are introduced and discussed. At this stage it will help to consider the notion of a *neighborhood* of a point.

DEFINITION A set Q of real numbers is a *neighborhood* of a real number x iff Q contains an interval of positive length centered at x—that is, iff there is $\epsilon > 0$ such that $(x - \epsilon, x + \epsilon) \subset Q$.

In particular, for each $\epsilon > 0$, $(x - \epsilon, x + \epsilon)$ is a neighborhood of x. In this setting, a sequence converges to A iff each neighborhood of A contains all but a finite number of terms of the sequence.

It is appropriate to state this as an unnumbered lemma and supply a proof. Although the facts are of an obvious nature, the proof will further an understanding of the definitions of convergence and a neighborhood.

LEMMA A sequence $\{a_n\}_{n=1}^{\infty}$ converges to A iff each neighborhood of A contains all but a finite number of terms of the sequence.

Proof. Suppose $\{a_n\}_{n=1}^{\infty}$ converges to A and let Q be a neighborhood of A. Then there is $\epsilon > 0$ such that $(A - \epsilon, A + \epsilon) \subset Q$. Since $\{a_n\}_{n=1}^{\infty}$ converges to A and $\epsilon > 0$, there is a positive integer N such that for $n \geqslant N$, $|a_n - A| < \epsilon$. In particular, for $n \geqslant N$, we have $A - \epsilon < a_n < A + \epsilon$; hence $a_n \in (A - \epsilon, A + \epsilon) \subset Q$. Thus Q contains all terms of the sequence except possibly some of the terms $a_1, a_2, \ldots, a_{N-1}$; hence Q contains all but a finite number of terms of the sequence.

Conversely, suppose each neighborhood of A contains all but a finite number of terms of the sequence $\{a_n\}_{n=1}^{\infty}$. Choose $\epsilon > 0$. Then $Q = (A - \epsilon, A + \epsilon)$ is a neighborhood of A and contains all but a finite number of terms of the sequence. This means that there is a finite set $S = \{n_1, n_2, \ldots, n_r\}$ of positive integers such that, if $a_n \notin Q$, then $n \in S$. Let $N = (\max S) + 1$. Now, if $n \geqslant N$, then $n \notin S$ and, by the definition of S, $a_n \in Q$; that is, $A - \epsilon < a_n < A + \epsilon$. This last statement is equivalent to the statement $|a_n - A| < \epsilon$. Hence, for $n \geqslant N$, $|a_n - A| < \epsilon$. We have shown that $\{a_n\}_{n=1}^{\infty}$ converges to A.

We now can answer a question that should be lurking in the back of your mind. Can a sequence $\{a_n\}_{n=1}^{\infty}$ converge to two different real numbers? Suppose the answer is yes. Then there would be distinct real numbers A and B such that $\{a_n\}_{n=1}^{\infty}$ converges to both A and B. Relying on our mental picture of the real line, we may easily convince ourselves that there are intervals P and Q, each of positive length, centered at A and B respectively, such that $P \cap Q$ is empty. You are asked to give a proof of this in Exercise 2. Now, as we have observed above, P contains all but a finite number of terms of the sequence $\{a_n\}_{n=1}^{\infty}$. Since $\{a_n\}_{n=1}^{\infty}$ also converges to B and since Q is a neighborhood of B, Q contains all but a finite number of terms of the sequence. Recalling that $P \cap Q$ is empty, we see that P must contain infinitely many terms of the sequence and also must contain at most a finite number of terms of the sequence. Thus, the assumption that $\{a_n\}_{n=1}^{\infty}$ converges to both A and B with $A \neq B$ leads to a contradiction. We shall state this as a theorem and give a careful proof. Note that our proof is only a restatement of the preceding argument in precise terms. It is worthwhile to point out at this stage of the game that this is the way a good share of mathematics is done. Theorems should follow naturally from definitions in an intuitive way; then, of course, adequate proofs must be given. At this level, a difficult theorem might be classified as one that does not seem to have an intuitive proof. Thus, we distinguish between a difficult theorem and a theorem that is difficult to prove even though the idea of its proof may be intuitively clear.

1.1 THEOREM If $\{a_n\}_{n=1}^{\infty}$ converges to A and also to B, then $A = B$.

Proof. Suppose $\{a_n\}_{n=1}^{\infty}$ converges to A and also to B with $A \neq B$, and we may as well suppose $A < B$. Let $\epsilon = \frac{1}{2}(B - A) > 0$. There is N such that $n \geqslant N$ implies that

$$A - \epsilon < a_n < A + \epsilon = A + \frac{1}{2}(B - A) = \frac{1}{2}(A + B).$$

Observe that $\frac{1}{2}(A + B) = B - \frac{1}{2}(B - A) = B - \epsilon$. Thus, the statement $B - \epsilon < a_n < B + \epsilon$ is false for all $n \geqslant N$, contrary to $\{a_n\}_{n=1}^{\infty}$ converging to B.

You may well be able to supply better proofs for this theorem. The proof here is patterned after our initial idea. This is not always the best method of proof, and we shall not adhere rigidly to this procedure.

Let us now consider the sequence $\{n\}_{n=1}^{\infty}$, or the sequence whose nth term is the positive integer n. It seems reasonably clear that this sequence cannot converge since the terms get larger and larger without bound. A sequence $\{a_n\}_{n=1}^{\infty}$ is *bounded from above* iff there is a real number M such that $a_n \leqslant M$ for all n. A sequence $\{a_n\}_{n=1}^{\infty}$ is *bounded from below* iff there is a real number P such that $P \leqslant a_n$ for all n. A sequence is *bounded* iff it is both bounded from above and bounded from below. Note that a sequence $\{a_n\}_{n=1}^{\infty}$ is bounded iff there are real numbers P and M such that $P \leqslant a_n \leqslant M$ for all n or, equivalently, iff there is a real number S such that $|a_n| \leqslant S$ for all n. Let us now suppose that $\{a_n\}_{n=1}^{\infty}$ converges to A. If we choose a neighborhood of A, there are, at most, a finite number of terms of the sequence outside this neighborhood. In particular, if we consider the neighborhood $(A - 1, A + 1)$, there is a positive integer N such that $A - 1 < a_n < A + 1$ for all $n \geq N$. Now we are assured that all terms of the sequence, except possibly the terms $a_1, a_2, \ldots, a_{N-1}$, are bounded from below by $A - 1$ and from above by $A + 1$. It is now easy to find upper and lower bounds for the possibly wayward terms $a_1, \ldots, a_{N-1}$, since this is a finite set of real numbers. It should now be clear how upper and lower bounds can be chosen for the sequence $\{a_n\}_{n=1}^{\infty}$.

1.2 THEOREM If $\{a_n\}_{n=1}^{\infty}$ converges to A, then $\{a_n\}_{n=1}^{\infty}$ is bounded.

Proof. Suppose $\{a_n\}_{n=1}^{\infty}$ converges to A. Choose $\epsilon = 1$. There is a positive integer N such that, if $n \geq N$, then $A - 1 < a_n < A + 1$. Let $S = \min\{a_1, a_2, \ldots, a_{N-1}, A - 1\}$ and let $M = \max\{a_1, a_2, \ldots, a_{N-1}, A + 1\}$. Then, for all n, we have $S \leq a_n \leq M$. Thus $\{a_n\}_{n=1}^{\infty}$ is bounded.

Note how easily it is to give a clear, concise proof for a theorem, such as the preceding one, once the idea of the proof has been laid out. The student of analysis should work to cultivate the skill of formulating proofs so as to communicate mathematics to fellow mathematicians in a clear, uncluttered style.

DEFINITION A sequence $\{a_n\}_{n=1}^{\infty}$ is said to be *convergent* iff there is a real number A such that $\{a_n\}_{n=1}^{\infty}$ converges to A. If $\{a_n\}_{n=1}^{\infty}$ is not convergent, it is said to be *divergent.*

In the light of this definition, let us analyze the facts to this point. If a sequence is convergent, then there is a real number A to which it converges, and, by Theorem 1.1, this number is unique. If a sequence is convergent, the unique number to which it converges is called the *limit* of the sequence.

In order to prove that a sequence $\{a_n\}_{n=1}^{\infty}$ is convergent, we must first guess to what real number it converges and then prove this conjecture. In many cases the determination of the supposed limit may be the more difficult task. Once the appropriate choice for the limit is found, the proof of convergence may be very easy.

■ **Example 1.1** Consider now the sequences $\left\{\frac{1}{n}\right\}_{n=1}^{\infty}$ and $\left\{1 + \frac{1}{n}\right\}_{n=1}^{\infty}$. If our intuition has been properly conditioned, we should recognize that the first sequence should converge to 0 and the second sequence should converge to 1. Let us prove both conjectures. Choose $\epsilon > 0$. There is a positive integer N such that $N > \frac{1}{\epsilon}$ or equivalently $\frac{1}{N} < \epsilon$. Thus, for $n \geq N$ we have

$$\left|\frac{1}{n} - 0\right| = \frac{1}{n} \leq \frac{1}{N} < \epsilon \text{ and } \left|\left(1 + \frac{1}{n}\right) - 1\right| = \frac{1}{n} \leq \frac{1}{N} < \epsilon.$$ ■

Note that we were able to use the same N for both sequences. This is not accidental. Observe that the nth term of the second sequence is obtained by adding 1 to the nth term of the first sequence. If one imagines the terms of the two sequences plotted on the real line, the two sequences appear to be in some sense "congruent," each term of the second sequence situated exactly one unit to the right of the corresponding term of the first sequence. It seems reasonable to say that the two sequences behave essentially the same except that they have different limits. We are led to conjecture that convergence is an internal property of a sequence and does not depend on what the limit happens to be. This is the topic we shall pursue next.

1.2 CAUCHY SEQUENCES

Suppose $\{a_n\}_{n=1}^{\infty}$ converges to A. Thus, the terms of the sequence must get close to A; in particular, if a_n and a_m are both close to A, then a_n and a_m must be close to each other. Let us formalize this notion.

DEFINITION A sequence $\{a_n\}_{n=1}^{\infty}$ is *Cauchy* iff for each $\epsilon > 0$ there is a positive integer N such that if $m, n \geqslant N$, then

$$|a_n - a_m| < \epsilon.$$

Let us consider our previous remarks. Suppose $\{a_n\}_{n=1}^{\infty}$ converges to A. Choose $\epsilon > 0$. There is a positive integer N such that, if $n, m \geqslant N$, then $A - \epsilon < a_n < A + \epsilon$ and $A - \epsilon < a_m < A + \epsilon$. Thus, for all $n, m \geqslant N$, we find $a_n \in (A - \epsilon, A + \epsilon)$ and $a_m \in (A - \epsilon, A + \epsilon)$. The set $(A - \epsilon, A + \epsilon)$ is an interval of length 2ϵ; hence, the difference between a_n and a_m is less than 2ϵ. We will now state a theorem, the proof of which we have just outlined.

1.3 THEOREM Every convergent sequence is a Cauchy sequence.

Proof. Suppose $\{a_n\}_{n=1}^{\infty}$ converges to A. Choose $\epsilon > 0$. Then $\frac{\epsilon}{2} > 0$. There is a positive integer N such that $n \geqslant N$ implies $|a_n - A| < \frac{\epsilon}{2}$. (The choice of $\frac{\epsilon}{2}$ is not a mere whim. We observed previously that the difference between a_n and a_m was less than twice the original choice of ϵ.) Now, if $m, n \geqslant N$, then $|a_n - A| < \frac{\epsilon}{2}$ and $|a_m - A| < \frac{\epsilon}{2}$; hence,

$$\begin{aligned}|a_n - a_m| &= |a_n - A + A - a_m| \leqslant |a_n - A| + |A - a_m| \\ &= |a_n - A| + |a_m - A| < \frac{\epsilon}{2} + \frac{\epsilon}{2} = \epsilon.\end{aligned}$$

Thus $\{a_n\}_{n=1}^{\infty}$ is Cauchy.

You should digest the basic idea rather than memorize the formal details of the proof. The manner of expressing the idea in a clear, precise fashion will come with experience.

Theorem 1.3 gives a necessary condition for convergence. If a sequence is convergent, it must be Cauchy. Equivalently, if a sequence is not Cauchy, then it is not convergent. Consider the sequences $\{1, 2, 3, \ldots\}$ and $\{1, 0, 1, 0, \ldots\}$. It is clear that neither is Cauchy; hence, according to Theorem 1.3, both fail to converge. As mentioned before, the property of being Cauchy is an internal property of a sequence. You might well suspect that this property is not only necessary but also sufficient for convergence. This is true, but a bit of work to prove. The next task will be to prove it.

Consider a Cauchy sequence $\{a_n\}_{n=1}^{\infty}$. Let $\epsilon = 1$. There is a positive integer N such that for $n, m \geqslant N$, $|a_n - a_m| < 1$. In particular, $N \geqslant N$. Hence, for $n \geqslant N$, $|a_N - a_n| < 1$; that is, $a_N - 1 < a_n < a_N + 1$. If you will recall the remarks preceding Theorem 1.2, it will be clear that every Cauchy sequence is bounded, and you should be able to give a nice proof of this theorem for Exercise 12.

1.4 THEOREM Every Cauchy sequence is bounded.

Proof. The proof is left as Exercise 12.

Consider again a Cauchy sequence $\{a_n\}_{n=1}^{\infty}$. The terms of this sequence all lie in an interval (C, D) since the sequence is bounded; moreover, the terms of the sequence get closer and closer together as one goes out farther and farther in the sequence. It seems reasonable to suspect that there is a real number A at which the terms of the sequence "pile up." We must find this number A and prove that the sequence converges to A. The existence of such a real number will be proved as a corollary to a theorem that we shall have cause to use later. First, the notion of a "pile up" must be formulated in precise terminology.

DEFINITION Let S be a set of real numbers. A real number A is an *accumulation point* of S iff every neighborhood of A contains infinitely many points of S.

A few remarks concerning this definition are in order. First of all, if A is an accumulation point of S, then every neighborhood of A contains at least one point of S that is different from A. (Indeed, A might not belong to S.) On the other hand, if A is not an accumulation point of S, then some neighborhood of A contains only a finite number of members of S. In this case it is possible to find a smaller neighborhood of A that excludes all points of S different from A. Thus, A is an accumulation point of S iff every neighborhood of A contains a member of S that is different from A. We state this result as follows.

LEMMA Let S be a set of real numbers. Then A is an accumulation point of S iff each neighborhood of A contains a member of S different from A.

■ **Example 1.2** Consider the set $S = \{\frac{1}{n} : n \text{ is a positive integer}\}$. This set is the range of the sequence $\{\frac{1}{n}\}_{n=1}^{\infty}$. Earlier, we showed that $\{\frac{1}{n}\}_{n=1}^{\infty}$ converges to zero. Thus, every neighborhood of 0 contains infinitely many terms of the sequence; and, since all terms of the sequence are distinct (that is, if $m \neq n$, then $a_m \neq a_n$), every neighborhood of 0 contains infinitely many points of the set S. Therefore, 0 is an accumulation point of the set S. ■

Observe that we did not define an accumulation point of a sequence but rather, an accumulation point of a set. One might be led to conjecture that the limit of a convergent sequence is always an accumulation point of the range of the sequence. To see that this is false, consider the sequence $\{a_n\}_{n=1}^{\infty}$ where $a_n = 1$ for all n. This sequence converges to 1, but its range is finite and hence, can have no accumulation points.

In Chapter 0, we noted that between any two real numbers there are infinitely many rational numbers and infinitely many irrational numbers. This yields the following theorem.

1.5 THEOREM Every real number is an accumulation point of the set of rational numbers. Every real number is an accumulation point of the set of irrational numbers.

Proof. Let x be any real number and Q be a neighborhood of x. There is $\epsilon > 0$ such that $(x - \epsilon, x + \epsilon) \subset Q$. There are infinitely many rational numbers between x and $x + \epsilon$, hence Q contains infinitely many rational numbers. Thus x is an accumulation point of the set of rational numbers. The proof for the set of irrational numbers is almost identical.

The class of finite sets of real numbers is a class of subsets of the real line, each of which does not possess accumulation points. A natural question to pursue is the following: Under what conditions can one guarantee that a set will have at least one accumulation point? Of course, such sets must be infinite. Now consider the set J of all positive integers. Given any real number A, it is easy to find a neighborhood of A that will exclude all but a finite number of positive integers (in fact, all if $A \not\subseteq J$). Thus J has no accumulation points. The defect here that allows J to avoid possessing accumulation points is that J has plenty of room to "spread out" its members. Perhaps if J is restricted so that it must be contained in an interval of finite length, this restriction will force the existence of accumulation points. This conjecture is true, as we shall now prove.

1.6 BOLZANO–WEIERSTRASS THEOREM Every bounded infinite set of real numbers has at least one accumulation point.

We shall preface the proof of this famous theorem with some comments on its meaning. As observed previously, it is impossible for finite sets to have accumulation points; moreover, we have an example—the set of natural numbers—of an infinite set with no accumulation point. Consequently, the boundedness of the set must play some role in forcing the existence of an accumulation point. Note that the theorem does not assert that the accumulation point need belong to the set.

Proof. Let S be a bounded infinite set. Since S is bounded, there are real numbers α and β such that $S \subset [\alpha, \beta]$. If α_1 is the midpoint of this interval, then at least one of the sets $[\alpha, \alpha_1]$ and $[\alpha_1, \beta]$ must contain an infinite set of members of S. Choose one with this property and call it $[a_1, b_1]$. If α_2 is the midpoint of this interval, then at least one of the sets $[a_1, \alpha_2]$ and $[\alpha_2, b_1]$ must contain an infinite set of members of S. Choose one with this property and call it $[a_2, b_2]$. Continuing in this fashion (actually, we are constructing two sequences, $\{a_n\}_{n=1}^{\infty}$ and $\{b_n\}_{n=1}^{\infty}$, by induction), we obtain, for each positive integer n, a closed interval $[a_n, b_n]$ with the following properties:

i. $b_n - a_n = 2^{-n}(\beta - \alpha)$.
ii. $[a_n, b_n]$ contains infinitely many points of S.
iii. $[a_n, b_n] \subset [a_{n-1}, b_{n-1}] \subset \cdots \subset [a_1, b_1] \subset [\alpha, \beta]$.

Since $[a_n, b_n] \subset [\alpha, \beta]$ for all n, the set $Q = \{a_n : n = 1, 2, \ldots\}$ is bounded; let $t = \sup Q$. This number will turn out to be the desired accumulation point. Recall that to prove this we need only show that every neighborhood of t contains infinitely many points of S. Let P be any neighborhood of t; then there is $\epsilon > 0$ such that $(t - \epsilon, t + \epsilon) \subset P$. Now $t - \epsilon$ is not an upper bound for Q, since t is the least upper bound; hence, there is a positive integer n such that $t - \epsilon < a_n \leqslant t$. In fact, if $m > n$, then, by the construction above, $t - \epsilon < a_n \leqslant a_m \leqslant t$. Each interval $[a_m, b_m]$ contains infinitely many points of S; hence, the proof would be complete if m could be found such that $t - \epsilon < a_m < b_m < t + \epsilon$. As noted above, $t - \epsilon < a_m \leqslant t$ for $m \geqslant n$; hence, it will suffice to choose m large enough so that $m \geqslant n$ and $2^{-m}(\beta - \alpha) < \epsilon$. ($2^{-m}(\beta - \alpha)$ is the length of the interval $[a_m, b_m]$.) To summarize, choose $m \geqslant n$ such that $2^{-m}(\beta - \alpha) < \epsilon$; then $t - \epsilon < a_m \leqslant t \leqslant b_m = a_m + 2^{-m}(\beta - \alpha) < t + \epsilon$. Thus P contains $[a_m, b_m]$; hence, P contains infinitely many members of S, and t is an accumulation point of S.

Suppose now that $\{a_n\}_{n=1}^{\infty}$ is a Cauchy sequence. If the range is finite, say $\{s_1, \ldots, s_r\}$. and if we choose

$$\epsilon = \min \{|s_i - s_j| : i \neq j, i, j = 1, \ldots, r\},$$

then there is a positive integer N such that $n, m \geqslant N$ implies that $|a_n - a_m| < \epsilon$. Since $a_n = s_j$ and $a_m = s_k$ for some j and k among $\{1, \ldots, r\}$ and ϵ was chosen to be the least distance between distinct members of the range, we must have $a_n = a_m$ for $n, m \geq N$. Thus, the sequence is constant from some point on and hence converges (see Exercise 11).

If the range of a Cauchy sequence is infinite, then, by Theorem 1.4, the range is an infinite bounded set and hence, by the Bolzano–Weierstrass Theorem, has an accumulation point. We have been searching for a point where the sequence "piles up," and this accumulation point should be it.

1.7 THEOREM Every Cauchy sequence is convergent.

Proof. Let $\{a_n\}_{n=1}^{\infty}$ be a Cauchy sequence. By the remarks above, if the range is finite, the sequence is constant from some point on; hence it converges. Suppose the range is infinite—call it S. By Theorem 1.4, S is bounded; hence, by the Bolzano–Weierstrass Theorem, S has an accumulation point—call it a. We shall prove that $\{a_n\}_{n=1}^{\infty}$ converges to a. Choose $\epsilon > 0$. Since $\left(a - \frac{\epsilon}{2}, a + \frac{\epsilon}{2}\right)$ is a neighborhood of a, it contains infinitely many members of the set S. Since $\{a_n\}_{n=1}^{\infty}$ is Cauchy, there is a positive integer N such that $n, m \geq N$ implies $|a_n - a_m| < \frac{\epsilon}{2}$. Also, since $\left(a - \frac{\epsilon}{2}, a + \frac{\epsilon}{2}\right)$ contains infinitely many points of s, and hence infinitely many terms of the sequence $\{a_n\}_{n=1}^{\infty}$, there is $n_0 \geq N$ such that $a_{n_0} \in \left(a - \frac{\epsilon}{2}, a + \frac{\epsilon}{2}\right)$. Now if $n \geq n_0$, we have

$$|a_n - a| \leq |a_n - a_{n_0}| + |a_{n_0} - a| < \frac{\epsilon}{2} + \frac{\epsilon}{2} = \epsilon.$$

Thus $\{a_n\}_{n=1}^{\infty}$ converges to a.

By combining Theorems 1.3 and 1.7, we see that a sequence is Cauchy iff it is convergent. It is now possible to recognize a convergent sequence without having any idea what the limit might be. This is not an unpleasant state of affairs, as we shall see later.

1.3 ARITHMETIC OPERATIONS ON SEQUENCES

We have been sampling the topology of the real line when convenient for our purposes, and our experiences with real numbers have been based mainly on the arithmetic processes—addition, subtraction, multiplication, and division—and the ordering of the real numbers. It is not surprising that these operations and the order relation are "topologically nice." The true meaning of this last sentence will become clearer in following chapters. For the present we shall be content to show the relationship between the arithmetic processes and the ordering on R and the notion of convergence.

Suppose $\{a_n\}_{n=1}^{\infty}$ and $\{b_n\}_{n=1}^{\infty}$ are Cauchy sequences. Considering the sequence $\{a_n + b_n\}_{n=1}^{\infty}$, we observe that the following inequality holds:

$$|(a_n + b_n) - (a_m + b_m)| = |(a_n - a_m) + (b_n - b_m)| \leq |a_n - a_m| + |b_n - b_m|.$$

The reader should now be able to see how to prove that the sequence $\{a_n + b_n\}_{n=1}^{\infty}$ is Cauchy. Thus, if $\{a_n\}_{n=1}^{\infty}$ and $\{b_n\}_{n=1}^{\infty}$ are convergent, so is $\{a_n + b_n\}_{n=1}^{\infty}$. We wish, however, to prove a more informative theorem, which not only states that $\{a_n + b_n\}_{n=1}^{\infty}$ converges, but also asserts what the limit will be.

1.8 THEOREM If $\{a_n\}_{n=1}^{\infty}$ converges to A and $\{b_n\}_{n=1}^{\infty}$ converges to B, then $\{a_n + b_n\}_{n=1}^{\infty}$ converges to $A + B$.

Proof. Choose $\epsilon > 0$. There is a positive integer N_1 such that, if $n \geq N_1$, then $|a_n - A| < \frac{\epsilon}{2}$. In like manner, there is a positive integer N_2 such that $n \geq N_2$ implies $|b_n - B| < \frac{\epsilon}{2}$. Let $N = \max\{N_1, N_2\}$. Then, if $n \geq N$, we have $n \geq N_1$, so $|a_n - A| < \frac{\epsilon}{2}$ and $n \geq N_2$; hence $|b_n - B| < \frac{\epsilon}{2}$. Thus,

$$\begin{aligned} |(a_n + b_n) - (A + B)| &= |(a_n - A) + (b_n - B)| \\ &\leq |a_n - A| + |b_n - B| < \frac{\epsilon}{2} + \frac{\epsilon}{2} = \epsilon. \end{aligned}$$

Thus $\{a_n + b_n\}_{n=1}^{\infty}$ converges to $A + B$.

Proofs like the preceding do not stem from divine inspiration. You begin with what is needed, $|(a_n + b_n) - (A + B)| < \epsilon$, and then you work backward to find the appropriate choice of N. Although proofs are not written this way, they are conceived in this manner. Let us follow this process for the product $\{a_nb_n\}_{n=1}^{\infty}$ of two sequences to see how this process works.

Suppose $\{a_n\}_{n=1}^{\infty}$ converges to A and $\{b_n\}_{n=1}^{\infty}$ converges to B. In the light of previous discussion, it seems reasonable to try to prove that $\{a_nb_n\}_{n=1}^{\infty}$ converges to AB. Now for the scratch work:

$$\begin{aligned} |a_nb_n - AB| &= |a_nb_n - a_nB + a_nB - AB| = |a_n(b_n - B) + B(a_n - A)| \\ &\leq |a_n|\,|b_n - B| + |B|\,|a_n - A|. \end{aligned}$$

Since $\{a_n\}_{n=1}^{\infty}$ converges to A and $\{b_n\}_{n=1}^{\infty}$ converges to B, we can make $|a_n - A|$ and $|b_n - B|$ small. The constant $|B|$ poses no problem, but $|a_n|$ depends on the choice of n. Now it is appropriate to recall an earlier theorem—namely, that every convergent sequence is bounded. Thus, although $|a_n|$ depends on the choice of n, it can't be very large. Suppose $|a_n| \leq M$ for all n. Then, if $\epsilon > 0$ is chosen, we wish to force the following:

$$|a_n|\,|b_n - B| + |B|\,|a_n - A| \leq M\,|b_n - B| + |B|\,|a_n - A| < \epsilon.$$

This is satisfied if

$$|b_n - B| < \frac{\epsilon}{M + |B|} \quad \text{and} \quad |a_n - A| < \frac{\epsilon}{M + |B|}.$$

It now remains to formulate an elegant proof.

1.9 THEOREM If $\{a_n\}_{n=1}^{\infty}$ converges to A and $\{b_n\}_{n=1}^{\infty}$ converges to B, then $\{a_n b_n\}_{n=1}^{\infty}$ converges to AB.

Proof. Choose $\epsilon > 0$. Since $\{a_n\}_{n=1}^{\infty}$ is a convergent sequence, it is bounded; hence, there is a positive real number M such that $|a_n| \leq M$ for all n. Now $\epsilon' = \dfrac{\epsilon}{|B| + M} > 0$; hence, there is a positive integer N_1 such that $n \geq N_1$ implies $|a_n - A| < \epsilon'$ and a positive integer N_2 such that $n \geq N_2$ implies $|b_n - B| < \epsilon'$. Let $N = \max\{N_1, N_2\}$. For $n \geq N$,

$$\begin{aligned}
|a_n b_n - AB| &= |(a_n b_n - a_n B) + (a_n B - AB)| \\
&\leq |a_n|\,|b_n - B| + |B|\,|a_n - A| \\
&\leq M\,|b_n - B| + |B|\,|a_n - A| < M\epsilon' + |B|\epsilon' \\
&= \epsilon.
\end{aligned}$$

Thus $\{a_n b_n\}_{n=1}^{\infty}$ converges to AB.

Some facts follow immediately from Theorems 1.8 and 1.9. If $\{a_n\}_{n=1}^{\infty}$ converges to A and α is any real number, then the constant sequence $\{\alpha_n\}_{n=1}^{\infty}$ converges to α; hence, the sequence $\{\alpha a_n\}_{n=1}^{\infty}$ converges to αA. In particular, if $\alpha = -1$, $\{-a_n\}_{n=1}^{\infty}$ converges to $-A$. Combining this observation with Theorem 1.8, we see that, if $\{a_n\}_{n=1}^{\infty}$ converges to A and $\{b_n\}_{n=1}^{\infty}$ converges to B, then $\{a_n - b_n\}_{n=1}^{\infty}$ converges to $A - B$. More generally, if α and β are real numbers, then $\{\alpha a_n + \beta b_n\}_{n=1}^{\infty}$ converges to $\alpha A + \beta B$.

Be careful not to read any unintended meaning into Theorems 1.8 and 1.9. If $\{a_n\}_{n=1}^{\infty}$ and $\{b_n\}_{n=1}^{\infty}$ are sequences that do not converge, Theorem 1.8 does not assert that $\{a_n + b_n\}_{n=1}^{\infty}$ fails to converge. In the same way, Theorem 1.9 does not assert that $\{a_n b_n\}_{n=1}^{\infty}$ fails to converge. Exercises 23, 24, and 25 illustrate this point.

The problem with division must be approached with some caution. First of all, consider $\{a_n\}_{n=1}^{\infty}$ converging to A and $\{b_n\}_{n=1}^{\infty}$ converging to B. We wish to consider the sequence $\left\{\dfrac{a_n}{b_n}\right\}_{n=1}^{\infty}$ and might be led to conjecture that this converges to $\dfrac{A}{B}$. For this even to make sense, we must insist that $B \neq 0$ and that $b_n \neq 0$ for all n. Let us proceed with these facts in mind.

$$\left|\frac{a_n}{b_n} - \frac{A}{B}\right| = \left|\frac{a_nB - b_nA}{b_nB}\right| = \left|\frac{a_nB - AB + AB - b_nA}{b_nB}\right|$$
$$< \frac{|a_n - A|}{|b_n|} + \frac{|A|\,|B - b_n|}{|b_n|\,|B|}.$$

Considering this result in the same fashion as that preceding Theorem 1.9, we see that the quantities $|a_n - A|$ and $|B - b_n|$ can be made small and that the constant $\frac{|A|}{|B|}$ poses no difficulty. However, the factor $\frac{1}{|b_n|}$ must be bounded somehow; indeed, it suffices to keep $|b_n|$ away from zero. However, since $\{b_n\}_{n=1}^{\infty}$ converges to $B \neq 0$, this should not be difficult. Since $B \neq 0$, $\epsilon = \frac{|B|}{2} > 0$. The set $\left(B - \frac{\epsilon}{2}, B + \frac{\epsilon}{2}\right)$ contains all but a finite number of terms of the sequence $\{b_n\}_{n=1}^{\infty}$; hence, the choice of ϵ makes it clear that all terms in this neighborhood are "bounded away from zero." In other words, there is a positive number M such that $|b_n| \geq M$ for all but a finite number of terms of the sequence. We shall now prove this in a precise and economical fashion.

1.10 LEMMA If $\{b_n\}_{n=1}^{\infty}$ converges to B and $B \neq 0$, then there is a positive real number M and a positive integer N such that, if $n \geq N$, then $|b_n| \geq M$.

Proof. Since $B \neq 0$, $\frac{|B|}{2} = \epsilon > 0$. There is N such that, if $n \geq N$, then $|b_n - B| < \epsilon$. Let $M = \frac{|B|}{2}$. Thus for $n \geq N$,

$$|b_n| = |b_n - B + B| \geq |B| - |b_n - B| \geq |B| - \frac{|B|}{2} = \frac{|B|}{2} = M.$$

Note that, although the idea of this proof is fairly simple, the inequalities used in the proof are a bit tricky. It is tempting to yield to the inclination to give neat, economical proofs like this. In fact, you should try your hand at this pleasant pastime as you gain more confidence and experience and begin to be critical of the proofs appearing in this book.

In order to cast aside the veil obscuring the important facts in this proof, we shall give an alternate proof for the case when $B > 0$. The case $B < 0$ may be handled similarly.

Of course, this is more cumbersome than the first proof, but it is much easier to digest. If $B > 0$, then $\frac{B}{2} > 0$. Now there is a positive integer N such that, for $n \geq N$, $\frac{B}{2} = B - \frac{B}{2} < b_n < B + \frac{B}{2}$. In the first proof for this lemma, the two cases $B > 0$ and $B < 0$ were handled simultaneously by judicious use of inequalities concerning absolute values. The reader will learn a lot by writing out the proof when $B < 0$ and comparing the two methods of proof.

Equipped with this lemma, we are now ready to state and prove a theorem concerning the quotient of two sequences.

1.11 THEOREM If $\{a_n\}_{n=1}^{\infty}$ converges to A and $\{b_n\}_{n=1}^{\infty}$ converges to B, with $B \neq 0$ and $b_n \neq 0$ for all n, then $\left\{\dfrac{a_n}{b_n}\right\}_{n=1}^{\infty}$ converges to $\dfrac{A}{B}$.

Proof. Choose $\epsilon > 0$. By Lemma 1.10, there is a positive real number M and a positive integer N_1 such that $|b_n| \geqslant M$ for all $n \geqslant N_1$. Then

$$\epsilon' = \frac{M\epsilon}{1 + \left|\dfrac{A}{B}\right|} > 0.$$

(See the paragraphs preceding Lemma 1.10 to understand the reason for this choice of ϵ'.) There is a positive integer N_2 such that, for $n \geqslant N_2$, $|a_n - A| < \epsilon'$ and a positive integer N_3 such that $n \geqslant N_3$ implies $|b_n - B| < \epsilon'$. Let $N = \max\{N_1, N_2, N_3\}$. For $n \geqslant N$, $|b_n - B| < \epsilon'$, $|a_n - A| < \epsilon'$, and $|b_n| \geqslant M$. Thus,

$$\begin{aligned}
\left|\frac{a_n}{b_n} - \frac{A}{B}\right| &= \left|\frac{a_nB - b_nA}{b_nB}\right| = \left|\frac{a_nB - AB + AB - b_nA}{b_nB}\right| \\
&\leqslant \left|\frac{a_n - A}{b_n}\right| + \frac{|A|\,|b_n - B|}{|b_n|\,|B|} \\
&< \epsilon'\left(\frac{1}{|b_n|} + \frac{|A|}{|B|\,|b_n|}\right) \leqslant \epsilon'\frac{1}{M}\left[1 + \frac{|A|}{|B|}\right] = \epsilon.
\end{aligned}$$

You have now been initiated into the exclusive club of epsilon pickers. Since the secrets revealed are representatives of the types of epsilon-picking problems to be encountered later in this book, you should digest them carefully.

As a corollary to Theorem 1.11, observe that, if $\{b_n\}_{n=1}^{\infty}$ converges to B with $B \neq 0$ and $b_n \neq 0$ for all n, then $\{\frac{1}{b}\}_{n=1}^{\infty}$ converges to $\frac{1}{B}$. (Let $a_n = 1$ for all n and apply Theorem 1.11.)

■ **Example 1.3** Consider the sequence

$$\left\{\frac{n^3 - 100n^2 + n - 58}{2n^3 + 65n + 69}\right\}_{n=1}^{\infty}.$$

At first glance it would appear that there is little use in trying to apply Theorem 1.11 to this sequence since the sequences

$$\{n^3 - 100n^2 + n - 58\}_{n=1}^{\infty} \quad \text{and} \quad \{2n^3 + 65n + 69\}_{n=1}^{\infty}$$

are both unbounded and, hence, divergent. However, some slight adjustments may improve the situation. Now, for each positive integer n,

$$\frac{n^3 - 100n^2 + n - 58}{2n^3 + 65n + 69} = \frac{1 - 100\frac{1}{n} + \frac{1}{n^2} - 58\frac{1}{n^3}}{2 + 65\frac{1}{n^2} + 69\frac{1}{n^3}}.$$

We have already shown that $\left\{\frac{1}{n}\right\}_{n=1}^{\infty}$ converges to 0; hence, by Theorem 1.9, $\left\{\frac{1}{n^2}\right\}_{n=1}^{\infty}$ and $\left\{\frac{1}{n^3}\right\}_{n=1}^{\infty}$ both converge to 0. If we recall the remarks following Theorem 1.9, it is clear that $\left\{1 - 100\left(\frac{1}{n}\right) + \frac{1}{n^2} - 58\left(\frac{1}{n^3}\right)\right\}_{n=1}^{\infty}$ converges to

$$1 - (100)(0) + 0 - (58)(0) = 1,$$

and $\left\{2 + 65\left(\frac{1}{n^2}\right) + 69\left(\frac{1}{n^3}\right)\right\}_{n=1}^{\infty}$ converges to

$$2 + (65)(0) + 69(0) = 2.$$

Now, by Theorem 1.11,

$$\left\{\frac{n^3 - 100n^2 + n - 58}{2n^3 + 65n + 69}\right\}_{n=1}^{\infty} = \left\{\frac{1 - 100\frac{1}{n} + \frac{1}{n^2} - 58\frac{1}{n^3}}{2 + 65\frac{1}{n^2} + 69\frac{1}{n^3}}\right\}_{n=1}^{\infty}$$

converges to $\frac{1}{2}$. ■

■ **Example 1.4** Consider the sequence $\{\sqrt{n+1} - \sqrt{n}\}_{n=1}^{\infty}$. As we saw in Example 1.3, the form in which this sequence is given does not lend itself to the use of tools at our disposal, but a familiar algebraic device will remedy this problem. Note that, for $n > 0$,

$$\sqrt{n+1} - \sqrt{n} = \frac{n + 1 - n}{\sqrt{n+1} + \sqrt{n}} = \frac{1}{\sqrt{n+1} + \sqrt{n}}.$$

Now let us refer to some of the homework exercises rather than attack this problem directly. By exercise 26, $\left\{\frac{1}{\sqrt{n}}\right\}_{n=1}^{\infty}$ converges to 0, since $\left\{\frac{1}{n}\right\}_{n=1}^{\infty}$ converges to 0. Now

$$0 < \frac{1}{\sqrt{n+1} + \sqrt{n}} \leqslant \frac{1}{2\sqrt{n}},$$

and the sequence $\left\{\frac{1}{2\sqrt{n}}\right\}_{n=1}^{\infty}$ converges to 0; hence the sequence

$$\{\sqrt{n+1} - \sqrt{n}\}_{n=1}^{\infty} = \left\{\frac{1}{\sqrt{n+1} + \sqrt{n}}\right\}_{n=1}^{\infty}$$

converges to 0 by Exercise 9. ■

A relation between order and convergence remains to be shown. We have a specific relation in mind and shall pursue that exclusively. Suppose $\{a_n\}_{n=1}^{\infty}$ and $\{b_n\}_{n=1}^{\infty}$ are sequences converging to A and B, respectively. Suppose further that $a_n \leqslant b_n$ for all n. By this time it should be evident what the conclusion should be—namely, that $A \leqslant B$. However, any attempt to improve this result, by asserting that $a_n < b_n$ for all n implies $A < B$, fails. To see this, consider the sequences $\left\{\frac{1}{2n^2}\right\}_{n=1}^{\infty}$ and $\left\{\frac{1}{n}\right\}_{n=1}^{\infty}$. We have $\frac{1}{2n^2} < \frac{1}{n}$ for all n, but both sequences converge to 0. The anticipated theorem will be proved by assuming that $A > B$ and showing that this leads to a contradiction of the hypothesis. If $A > B$, then it is easy to find neighborhoods P and Q of A and B, respectively, that do not overlap and such that if $x \in P$ and $y \in Q$, then $x > y$. All but a finite number of terms of $\{a_n\}_{n=1}^{\infty}$ belong to the set P. The corresponding terms of the sequence $\{b_n\}_{n=1}^{\infty}$ cannot belong to Q, since $a_n \leqslant b_n$ for all n. Thus, there are at most a finite number of terms of the sequence $\{b_n\}_{n=1}^{\infty}$ belonging to the set Q contrary to $\{b_n\}_{n=1}^{\infty}$ converging to B.

1.12 THEOREM If $\{a_n\}_{n=1}^{\infty}$ converges to A and $\{b_n\}_{n=1}^{\infty}$ converges to B, with $a_n \leqslant b_n$ for all n, then $A \leqslant B$.

Proof. Suppose $B < A$. Then $\epsilon = \dfrac{A - B}{2} > 0$. There is a positive integer N_1 such that $n \geqslant N_1$ implies $A - \epsilon < a_n < A + \epsilon$. There is a positive integer N_2 such that $n \geqslant N_2$ implies $B - \epsilon < b_n < B + \epsilon$. Choose $N \geqslant \max\{N_1, N_2\}$. Then, for $n \geqslant N$,

$$b_n < B + \epsilon = A - \epsilon < a_n \leqslant b_n.$$

This last inequality is clearly impossible.

Note that the contradiction in the proof was not the one mentioned preceding the theorem. But the idea is essentially the same; the choice was motivated by a desire to make the proof simpler to present.

A word of friendly advice may help. Contrary to popular belief, mathematics does not consist entirely of formulas and recipes for solving problems. Some of them are convenient in mathematics, but, unless they are applied with an understanding of the underlying principles and some intuitive feeling for these principles, the rules and formulas have limited use. You should construct examples other than those given here and examine them critically. The ideas used in formulating a proof may be very useful in solving a problem in which the theorem may not apply. The next example illustrates this point.

Consider the sequence

$$\left\{\frac{\sin \frac{n\pi}{2}}{n}\right\}_{n=1}^{\infty}.$$

This can be viewed as the product of two sequences $\left\{\frac{1}{n}\right\}_{n=1}^{\infty}$ and $\left\{\sin \frac{n\pi}{2}\right\}_{n=1}^{\infty}$. Unfortunately, the sequence $\left\{\sin \frac{n\pi}{2}\right\}_{n=1}^{\infty}$ does not converge since $\sin \frac{n\pi}{2} = 0$ if n is divisible by 2, and $\sin \frac{n\pi}{2}$ is $+1$ or -1 if n is odd; thus, we cannot apply Theorem 1.9. However, $\left\{\sin \frac{n\pi}{2}\right\}_{n=1}^{\infty}$ is bounded, and $\left\{\frac{1}{n}\right\}_{n=1}^{\infty}$ converges to 0. Referring to the paragraph preceding Theorem 1.9, let $b_n = \frac{1}{n}$ and $a_n = \sin \frac{n\pi}{2}$. Then $B = 0$, and A is unimportant since $|B|\,|a_n - A| = 0$. Thus, it appears that the sequence converges to 0. Let's check this to make sure. Choose $\epsilon > 0$. There is a positive integer N such that $\frac{1}{N} < \epsilon$. Then, for $n \geqslant N$, we have

$$\left|\frac{\sin \frac{n\pi}{2}}{n}\right| \leqslant \frac{1}{n} \leqslant \frac{1}{N} < \epsilon.$$

Our conjecture was correct; in fact, we are led to invent a new theorem. Two facts were used in the discussion—namely, $\left\{\frac{1}{n}\right\}_{n=1}^{\infty}$ converges to 0, and $\left\{\sin \frac{n\pi}{2}\right\}_{n=1}^{\infty}$ is bounded.

1.13 THEOREM If $\{a_n\}_{n=1}^{\infty}$ converges to 0 and $\{b_n\}_{n=1}^{\infty}$ is bounded, then $\{a_n b_n\}_{n=1}^{\infty}$ converges to 0.

Proof. Let M be a positive number such that $|b_n| \leqslant M$ for all n. Choose $\epsilon > 0$. Then $\epsilon' = \frac{\epsilon}{M} > 0$. (The reason for this choice of ϵ' will be apparent shortly, but you should have anticipated this choice.) There is a positive integer N such that $n \geqslant N$ implies $|a_n| = |a_n - 0| < \epsilon'$. Then

$$|a_n b_n - 0| = |a_n b_n| = |a_n|\,|b_n| \leqslant |a_n|\,M < \epsilon' M = \epsilon.$$

Thus $\{a_n b_n\}_{n=1}^{\infty}$ converges to 0.

Now reconsider one of the equivalent formulations of the idea of convergence: a sequence $\{a_n\}_{n=1}^{\infty}$ converges to A iff each neighborhood of A contains all but a finite number of terms of the sequence. This statement makes it clear that any finite number of terms of a sequence may be changed without affecting either the convergence of the sequence or the limit; in fact, a finite number of terms may be deleted from the sequence and the remaining terms relabeled, retaining the original order, without affecting the convergence or limit. The definition of a subsequence of a sequence will facilitate this discussion.

1.4 SUBSEQUENCES AND MONOTONE SEQUENCES

DEFINITION Let $\{a_n\}_{n=1}^{\infty}$ be a sequence and $\{n_k\}_{k=1}^{\infty}$ be any sequence of positive integers such that $n_1 < n_2 < n_3 \cdots$. The sequence $\{a_{n_k}\}_{k=1}^{\infty}$ is called a *subsequence* of $\{a_n\}_{n=1}^{\infty}$.

In less formal terms, we can say that a subsequence is formed by deleting some (or none, since every sequence is a subsequence of itself) of the terms of the sequence and relabeling the remaining terms while retaining the original order of the terms. Of course, there must be infinitely many terms remaining after the deletion. For example, the sequences $\left\{\frac{1}{k^2}\right\}_{k=1}^{\infty}$, $\left\{\frac{1}{2k}\right\}_{k=1}^{\infty}$, and $\left\{\frac{1}{2^k}\right\}_{k=1}^{\infty}$ are subsequences of the sequence $\left\{\frac{1}{n}\right\}_{n=1}^{\infty}$ formed by setting $n_k = k^2$, $n_k = 2k$, and $n_k = 2^k$, respectively. The sequences $\left\{\frac{1}{k^2}\right\}_{k=1}^{\infty}$ and $\left\{\frac{1}{2k}\right\}_{k=1}^{\infty}$ converge to 0 according to our previous remarks. The reader is urged to try to prove that $\left\{\frac{1}{2^k}\right\}_{k=1}^{\infty}$ also converges to 0, although this fact will be evident from some results that follow.

If $a_n = \dfrac{1 + (-1)^n}{2}$ for each positive integer n and the sequence $\{n_k\}_{k=1}^{\infty}$ is chosen such that $n_k = 2k$ for each positive integer k, then $a_{n_k} = 1$ for each k, and the subsequence thus obtained has all terms equal to 1 and, hence, converges to 1. On the other hand, if n_k is chosen to be $2k - 1$ for each k, then $a_{n_k} = 0$ for all k, and the subsequence thus chosen converges to 0. As noted earlier, every sequence is a subsequence of itself, and this sequence does not converge.

Reflect for a moment on the two examples just given. The first was a convergent sequence, and, of the three subsequences sampled, all converged to the same limit: the limit of the original sequence. Although this evidence is not conclusive, it does give some "statistical support" to lead one to conjecture that every subsequence of a convergent sequence converges and to the same limit.

This result should not be surprising in light of the discussion preceding the definition of a subsequence. Proceeding with great confidence in the inherent orderliness of mathematics, we consider the second example given—a sequence that does not converge. It has two subsequences with different limits and one subsequence that does not converge at all. By now you should have a clear understanding of how the statements of many theorems may be formulated by considering well-chosen examples. In fact, you should associate with each theorem of importance a stock of examples and counterexamples that illustrate both the necessity of the hypotheses and the strength of the conclusions. A theorem used in this fashion becomes a valued aid and tool to be cherished and exploited, not a tedious fact to be memorized.

1.14 THEOREM A sequence converges iff each of its subsequences converges. In fact, if every subsequence converges, then they all converge to the same limit.

Proof. The first half of this theorem is very easy to prove, because every sequence is a subsequence of itself. Indeed, if every subsequence of a sequence converges, then the sequence is convergent, since the sequence is included among the subsequences, all of which were assumed convergent.

Assume now that $\{a_n\}_{n=1}^{\infty}$ is a sequence that converges to A and $\{a_{n_k}\}_{k=1}^{\infty}$ is a subsequence. The proof of the theorem will be complete if it can be shown that $\{a_{n_k}\}_{k=1}^{\infty}$ converges to A. (Before going further, read again the paragraph preceding the definition of a subsequence.) Choose $\epsilon > 0$. Since $\{a_n\}_{n=1}^{\infty}$ converges to A, there is a positive integer N such that for $n \geqslant N$, $|a_n - A| < \epsilon$. Since $\{a_{n_k}\}_{k=1}^{\infty}$ is a subsequence of the given sequence $n_1 < n_2 < \cdots$; hence, $k \leqslant n_k$. Thus, for $k \geqslant N$, $n_k \geqslant N$; hence $|a_{n_k} - A| < \epsilon$. Therefore, $\{a_{n_k}\}_{k=1}^{\infty}$ converges to A. Note that, if every subsequence of a given sequence converges, then the sequence converges, and, by the argument just given, all subsequences converge to the same limit.

We can prove a slight variation on Theorem 1.14 if we assume that the sequence is bounded.

1.15 THEOREM Suppose $\{x_n\}_{n=1}^{\infty}$ is a bounded sequence. If all of its convergent subsequences have the same limit, then the sequence is convergent.

Proof. Assume that all of the convergent subsequences of $\{x_n\}_{n=1}^{\infty}$ have the same limit, call it x_0. Exercise 34 guarantees that there is at least one convergent subsequence.

If $\{x_n\}_{n=1}^{\infty}$ does not converge to x_0, then there is $\epsilon > 0$ such that for each k there is $n_k > k$ such that $|x_{n_k} - x_0| \geqslant \epsilon$. The sequence $\{x_{n_k}\}_{k=1}^{\infty}$ being bounded will have a convergent subsequence that will converge to something different from x_0. This contradicts the hypothesis.

DEFINITION A sequence $\{a_n\}_{n=1}^{\infty}$ is *increasing* iff $a_n \leqslant a_{n+1}$ for all positive integers n. A sequence $\{a_n\}_{n=1}^{\infty}$ is *decreasing* iff $a_n \geqslant a_{n+1}$ for all positive integers n. A sequence is *monotone* iff it is either increasing or decreasing.

1.16 THEOREM A monotone sequence is convergent iff it is bounded.

Proof. Suppose $\{a_n\}_{n=1}^{\infty}$ is a monotone sequence that is bounded; for definiteness, suppose the sequence is increasing. The set $\{a_n : n = 1, 2, \ldots\}$ is bounded, so let $s = \sup \{a_n : n = 1, 2, \ldots\}$. It will be shown that $\{a_n\}_{n=1}^{\infty}$ converges to s. Choose $\epsilon > 0$. Since s is the least upper bound of $\{a_n : n = 1, 2, \ldots\}$, $s - \epsilon$ is not an upper bound; hence, there is n_0 such that $s - \epsilon < a_{n_0}$. Now for $n \geqslant n_0$,

$$s - \epsilon < a_{n_0} \leqslant a_n \leqslant s < s + \epsilon;$$

hence $\{a_n\}_{n=1}^{\infty}$ converges to s. For the case in which $\{a_n\}_{n=1}^{\infty}$ is decreasing, let $s = \inf \{a_n : n = 1, 2, \ldots\}$, and the details are quite similar (see Exercise 35).

If the sequence is convergent, then, by Theorem 1.2 it is bounded.

In keeping with our philosophy about examining examples in order to illustrate theorems and show the need for the hypotheses, we shall consider several sequences. First, the sequence $\{n\}_{n=1}^{\infty}$ is indeed monotone, but it does not converge, since it is unbounded. The sequence $\{0, 1, 0, \ldots\}$ is bounded, but the theorem does not apply, because it is not monotone; indeed, it fails to converge.

■ **Example 1.5** Consider the sequence $\{s_n\}_{n=1}^{\infty}$ defined as follows:

$$s_1 = \sqrt{2}, \quad s_n = \sqrt{2 + \sqrt{s_{n-1}}} \quad \text{for } n = 2, 3, \ldots .$$

We shall prove that this sequence is increasing and bounded by 2 and hence, by the preceding theorem, is convergent. Since this sequence is defined recursively, we will use induction to prove the conjecture stated above. Clearly, $s_1 \leqslant 2$ and

$$s_2 = \sqrt{2 + \sqrt[4]{2}} \geqslant \sqrt{2} = s_1.$$

We wish to show that $s_n \leqslant 2$ and $s_{n+1} \geqslant s_n$ for all n; the preceding statement verifies the truth of this for $n = 1$. Assume the condition is satisfied for $n = r$. Then

$$s_{r+1} = \sqrt{2 + \sqrt{s_r}} \leqslant \sqrt{2 + \sqrt{2}} \leqslant \sqrt{2 + 2} = 2$$

and

$$s_{r+2} = \sqrt{2 + \sqrt{s_{r+1}}} \geqslant \sqrt{2 + \sqrt{s_r}} = s_{r+1};$$

hence, by the principle of induction, the inequalities hold for all n. By use of Theorem 1.15, we are assured that the sequence converges; call the limit L. Now the subsequence $\{s_{n-1}\}_{n=2}^{\infty}$ also converges to L, and since $s_n > 0$ for all n,

$\{\sqrt{s_{n-1}}\}_{n=2}^{\infty}$ converges to $\sqrt{L}$. Now observe that since $s_n = \sqrt{2 + \sqrt{s_{n-1}}}$, the sequence

$$\{s_n\}_{n=1}^{\infty} = \{\sqrt{2 + \sqrt{s_{n-1}}}\}_{n=1}^{\infty}$$

converges to $\sqrt{2 + \sqrt{L}}$, and also $\{s_n\}_{n=1}^{\infty}$ converges to L; thus, by the uniqueness of the limit of a sequence, $L = \sqrt{2 + \sqrt{L}}$. By eliminating the radicals, we see that L must be a root of the polynomial equation $L^4 - 4L^2 - L + 4 = 0$.

■

It is left as an exercise to consider the sequence $\{(1 + \frac{1}{n})^n\}_{n=1}^{\infty}$. This sequence may be shown to be increasing and bounded, and hence convergent. The limit happens to be *e,* the base for the natural logarithm—a number that plays a central role in calculus. See the project at the end of the chapter.

1.17 THEOREM Let E be a set of real numbers. Then x_0 is an accumulation point of E iff there is a sequence $\{x_n\}_{n=1}^{\infty}$ of members of E, each distinct from x_0, such that $\{x_n\}_{n=1}^{\infty}$ converges to x_0.

Proof. Let x_0 be an accumulation point of E. Then, for each positive integer n, there is a point $x_n \in E$ such that $0 < |x_n - x_0| < \dfrac{1}{n}$. [The set $\left(x_0 - \dfrac{1}{n}, x_0 + \dfrac{1}{n}\right)$ is a neighborhood of x_0, so it contains a member of E distinct from x_0.] It remains to be shown that $\{x_n\}_{n=1}^{\infty}$ converges to x_0. Choose $\epsilon > 0$. Then there is a positive integer N such that $\dfrac{1}{\epsilon} < N$; hence, for $n \geq N$, $|x_n - x_0| < \dfrac{1}{n} \leq \dfrac{1}{N} \leq \epsilon$. Thus, $\{x_n\}_{n=1}^{\infty}$ converges to x_0.

Suppose now there is a sequence $\{x_n\}_{n=1}^{\infty}$ of members of E, each distinct from x_0, that converges to x_0. Since every neighborhood of x_0 contains all but a finite number of terms of the sequence, every neighborhood of x_0 must contain at least one member of E that is distinct from x_0. Hence, x_0 is an accumulation point of E.

We close this chapter by examining several examples involving techniques that are of sufficient interest to be studied with care.

■ **Example 1.6** Consider a real number $0 < b < 1$ and form the sequence $\{b^n\}_{n=1}^{\infty}$. Observe that $b^{n-1} - b^n = b^{n-1}(1 - b) > 0$; hence, the sequence is decreasing. In order to show that this sequence converges, it suffices to show that it is bounded from below, an easy task since $b^n > 0$ for all natural numbers n. Although we know now that the sequence converges, we shall seek also to determine its limit. Now $\{b^n\}_{n=1}^{\infty}$ converges; call its limit L. And $\{b^{2n}\}_{n=1}^{\infty}$, being a subsequence of $\{b^n\}_{n=1}^{\infty}$, also converges to L. On the other hand, $\{b^{2n}\}_{n=1}^{\infty}$, which may be considered as the product of the sequence $\{b^n\}_{n=1}^{\infty}$ with itself, converges to L^2. By the uniqueness of the limit, $L^2 = L$, and so $L = 0$ or $L = 1$. Clearly, $L \neq 1$ since the sequence was decreasing and $b < 1$. Hence, $\{b^n\}_{n=1}^{\infty}$ converges to 0. This example is not intended to amaze the reader, but it illustrates a few techniques that are convenient to have at hand. ■

■ **Example 1.7** Consider $0 < c < 1$ and the sequence $\{\sqrt[n]{c}\}_{n=1}^{\infty}$. For all n,

$$\sqrt[n]{c} - \sqrt[n-1]{c} = \sqrt[n]{c}(1 - \sqrt[n(n-1)]{c}) > 0,$$

since $c < 1$. Hence the sequence is increasing, and clearly $\sqrt[n]{c} < 1$ for all n. Thus, the sequence converges; call the limit L. By Exercise 26, if $\{x_n\}_{n=1}^{\infty}$ converges to x_0, $x_n \geq 0$ for all n, then $\{\sqrt{x_n}\}_{n=1}^{\infty}$ converges to $\sqrt{x_0}$. Therefore, the sequence $\{\sqrt{\sqrt[n]{c}}\}_{n=1}^{\infty}$ converges to $\sqrt{L}$, but

$$\sqrt{\sqrt[n]{c}} = \sqrt[2n]{c},$$

and $\{\sqrt[2n]{c}\}_{n=1}^{\infty}$ is a subsequence of $\{\sqrt[n]{c}\}_{n=1}^{\infty}$ and hence converges to L. By the uniqueness of the limit of a sequence, $\sqrt{L} = L$; hence $L = 0$ or $L = 1$. But $c > 0$ and the sequence is increasing; therefore $L \neq 0$. In conclusion $\{\sqrt[n]{c}\}_{n=1}^{\infty}$ converges to 1. (See Exercise 36 for the case $c > 1$.) ■

■ **Example 1.8** We shall now consider another sequence that is defined recursively—that is, the first few terms are given, and instructions are given for computing the nth term in terms of some or all of the preceding terms. Let $a_1 = 1$, and for $n \geq 2$, define $a_n = \sqrt{2a_{n-1}}$. We shall also use induction to attack this sequence. First of all, look at a few terms of the sequence,

$$a_1 = 1, \quad a_2 = \sqrt{2}, \quad a_3 = \sqrt{2\sqrt{2}},$$

It seems reasonable to expect the sequence to be increasing and, hence, convergent if it is bounded. It may not be clear at first glance what to try for an upper bound; so we shall discover this bound by a very useful device. (Even if you can guess the bound, play the game to see how the device works.) Since the sequence is increasing, if it converges, the limit will be the least upper bound of the sequence. In an attempt to discover a logical candidate for this limit, let us assume that the sequence converges. Suppose $\{a_n\}_{n=1}^{\infty}$ converges to L; then $\{a_n^2\}_{n=1}^{\infty}$ converges to L^2, but $a_n^2 = 2a_{n-1}$; hence $\{a_n^2\}_{n=1}^{\infty} = \{2a_{n-1}\}_{n=1}^{\infty}$ converges to $2L$. Thus, the only candidates for the limit must be the solutions of the equation $L^2 = 2L$; either $L = 2$ or $L = 0$. It is easy to rule out the possibility $L = 0$, and so $L = 2$ is the obvious choice. Note that it has not been proved that $\{a_n\}_{n=1}^{\infty}$ converges to 2. Our only proof so far is that if the sequence converges, then the limit must be 2. We shall now prove by induction that $a_n \leq a_{n+1} \leq 2$ for all n; hence, the sequence is bounded and monotone and, thus, convergent. For $n = 1$, $a_1 = 1 \leq a_2 = \sqrt{2} \leq 2$. Assume the statement is true for $n = r$; in other words, $a_r \leq a_{r+1} \leq 2$. Then

$$a_{r+1} = \sqrt{2a_r} \leq \sqrt{2a_{r+1}} = a_{r+2}$$

and

$$a_{r+2} = \sqrt{2a_{r+1}} < \sqrt{2 \cdot 2} = 2$$

and the statement holds for $n = r + 1$. By induction, the statement holds for all n, and $\{a_n\}_{n=1}^{\infty}$ converges to 2. ■

■ **Example 1.9** Recall now a word of philosophy mentioned earlier—namely, that determining the limit of a sequence may be half the battle in showing a sequence to be convergent. Consider the sequence $\{\sqrt[n]{n}\}_{n=1}^{\infty}$. Let us try to guess the limit in advance by reasoning similar to that used in the preceding paragraph. Suppose $\{\sqrt[n]{n}\}_{n=1}^{\infty}$ converges; call the limit L. Now consider the subsequence $\{\sqrt[2n]{2n}\}_{n=1}^{\infty}$;

$$\sqrt[2n]{2n} = \sqrt{\sqrt[n]{2}\sqrt[n]{n}},$$

and we know $\{\sqrt[n]{2}\}_{n=1}^{\infty}$ converges to 1 (see Exercise 36). Thus, $\{\sqrt[2n]{2n}\}_{n=1}^{\infty}$ converges to L and also to $\sqrt{L}$; hence, by arguments given before, $L = 1$. We shall try to prove that the sequence converges to 1 or, equivalently, that the sequence $\{\sqrt[n]{n} - 1\}_{n=1}^{\infty}$ converges to 0. Let $x_n = \sqrt[n]{n} - 1$; clearly $x_n \geqslant 0$ and

$$n = (1 + x_n)^n = 1 + nx_n + \frac{n(n-1)}{2}x_n^2 + \cdots + x_n^n \geqslant \frac{n(n-1)}{2}x_n^2.$$

Thus, for all $n \geqslant 2$, we have $0 \leqslant x_n \leqslant \sqrt{\dfrac{2}{n-1}}$. It should be clear now how to complete the proof that $\{x_n\}_{n=1}^{\infty}$ converges to 0. ■

EXERCISES

1.1 SEQUENCES AND CONVERGENCE

1. Show that $[0, 1]$ is a neighborhood of $\frac{2}{3}$—that is, there is $\epsilon > 0$ such that $\left(\frac{2}{3} - \epsilon, \frac{2}{3} + \epsilon\right) \subset [0, 1]$.

*2. Let x and y be distinct real numbers. Prove that there is a neighborhood of P of x and a neighborhood Q of y such that $P \cap Q = \varnothing$.

*3. Suppose x is a real number and $\epsilon > 0$. Prove that $(x - \epsilon, x + \epsilon)$ is a neighborhood of each of its members; in other words, if $y \in (x - \epsilon, x + \epsilon)$, then there is $\delta > 0$ such that $(y - \delta, y + \delta) \subset (x - \epsilon, x + \epsilon)$.

4. Find upper and lower bounds for the sequence $\left\{\dfrac{3n + 7}{n}\right\}_{n=1}^{\infty}$.

5. Give an example of a sequence that is bounded but not convergent.

6. Use the definition of convergence to prove that each of the following sequences converges:

   a. $\left\{5 + \dfrac{1}{n}\right\}_{n=1}^{\infty}$ b. $\left\{\dfrac{2 - 2n}{n}\right\}_{n=1}^{\infty}$ c. $\left\{\dfrac{n + 1}{n}\right\}_{n=1}^{\infty}$ d. $\left\{\dfrac{3n}{2n + 1}\right\}_{n=1}^{\infty}$

***7.** Show that $\{a_n\}_{n=1}^{\infty}$ converges to A iff $\{a_n - A\}_{n=1}^{\infty}$ converges to 0.

8. Suppose $\{a_n\}_{n=1}^{\infty}$ converges to A and define a new sequence $\{b_n\}_{n=1}^{\infty}$ by $b_n = \dfrac{a_n + a_{n+1}}{2}$ for all n. Prove that $\{b_n\}_{n=1}^{\infty}$ converges to A.

***9.** Suppose $\{a_n\}_{n=1}^{\infty}$, $\{b_n\}_{n=1}^{\infty}$, and $\{c_n\}_{n=1}^{\infty}$ are sequences such that $\{a_n\}_{n=1}^{\infty}$ converges to A, $\{b_n\}_{n=1}^{\infty}$ converges to A, and $a_n \leqslant c_n \leqslant b_n$ for all n. Prove that $\{c_n\}_{n=1}^{\infty}$ converges to A.

***10.** Prove that, if $\{a_n\}_{n=1}^{\infty}$ converges to A, then $\{|a_n|\}_{n=1}^{\infty}$ converges to $|A|$. Is the converse true? Justify your conclusion.

***11.** Let $\{a_n\}_{n=1}^{\infty}$ be a sequence such that there exist numbers α and N such that, for $n \geqslant N$, $a_n = \alpha$. Prove that $\{a_n\}_{n=1}^{\infty}$ converges to α.

1.2 CAUCHY SEQUENCES

***12.** Prove that every Cauchy sequence is bounded (Theorem 1.4).

13. Prove directly (do not use Theorem 1.8) that, if $\{a_n\}_{n=1}^{\infty}$ and $\{b_n\}_{n=1}^{\infty}$ are Cauchy, so is $\{a_n + b_n\}_{n=1}^{\infty}$.

14. Prove directly (do not use Theorem 1.9) that, if $\{a_n\}_{n=1}^{\infty}$ and $\{b_n\}_{n=1}^{\infty}$ are Cauchy, so is $\{a_nb_n\}_{n=1}^{\infty}$. You will want to use Theorem 1.4.

15. Prove that the sequence $\left\{\dfrac{2n + 1}{n}\right\}_{n=1}^{\infty}$. is Cauchy.

16. Give an example of a set with exactly two accumulation points.

17. Give an example of a set with a countably infinite set of accumulation points.

18. Give an example of a set that contains each of its accumulation points.

19. Determine the accumulation points of the set $\left\{2^n + \dfrac{1}{k} : n \text{ and } k \text{ are positive integers}\right\}$

***20.** Let S be a nonempty set of real numbers that is bounded from above (below) and let $x = \sup S$ ($\inf S$). Prove that either x belongs to S or x is an accumulation point of S.

21. Let a_0 and a_1 be distinct real numbers. Define $a_n = \dfrac{a_{n-1} + a_{n-2}}{2}$ for each positive integer $n \geqslant 2$. Show that $\{a_n\}_{n=1}^{\infty}$ is a Cauchy sequence. You may want to use induction to show that $a_{n+1} - a_n = \left(-\dfrac{1}{2}\right)^n (a_1 - a_0)$ and then use the result from Example 0.9 of Chapter 0.

22. Suppose $\{a_n\}_{n=1}^{\infty}$ converges to A and $\{a_n : n \in J\}$ is an infinite set. Show that A is an accumulation point of $\{a_n : n \in J\}$.

1.3 ARITHMETIC OPERATIONS ON SEQUENCES

23. Suppose $\{a_n\}_{n=1}^{\infty}$ and $\{b_n\}_{n=1}^{\infty}$ are sequences such that $\{a_n\}_{n=1}^{\infty}$ and $\{a_n + b_n\}_{n=1}^{\infty}$ converge. Prove that $\{b_n\}_{n=1}^{\infty}$ converges.

24. Give an example in which $\{a_n\}_{n=1}^{\infty}$ and $\{b_n\}_{n=1}^{\infty}$ do not converge but $\{a_n + b_n\}_{n=1}^{\infty}$ converges.

25. Suppose $\{a_n\}_{n=1}^{\infty}$ and $\{b_n\}_{n=1}^{\infty}$ are sequences such that $\{a_n\}_{n=1}^{\infty}$ converges to $A \neq 0$ and $\{a_nb_n\}_{n=1}^{\infty}$ converges. Prove that $\{b_n\}_{n=1}^{\infty}$ converges.

***26.** If $\{a_n\}_{n=1}^{\infty}$ converges to a with $a_n \geq 0$ for all n, show $\{\sqrt{a_n}\}_{n=1}^{\infty}$ converges to $\sqrt{a}$. [*Hint:* If $a > 0$, then $\sqrt{a_n} - \sqrt{a} = \dfrac{a_n - a}{\sqrt{a_n} + \sqrt{a}}$.]

27. Prove that

$$\left\{\frac{\binom{n+k}{k}}{(n+k)^k}\right\}_{n=1}^{\infty}$$

converges to $\dfrac{1}{k!}$, where

$$\binom{n+k}{k} = \frac{(n+k)!}{n!k!}.$$

28. Prove the following variation on Lemma 1.10. If $\{b_n\}_{n=1}^{\infty}$ converges to $B \neq 0$ and $b_n \neq 0$ for all n, then there is $M > 0$ such that $|b_n| \geq M$ for all n.

29. Consider a sequence $\{a_n\}_{n=1}^{\infty}$ and, for each n, define

$$\alpha_n = \frac{a_1 + a_2 + \cdots + a_n}{n}.$$

Prove that if $\{a_n\}_{n=1}^{\infty}$ converges to A, then $\{\alpha_n\}_{n=1}^{\infty}$ converges to A. Give an example in which $\{\alpha_n\}_{n=1}^{\infty}$ converges, but $\{a_n\}_{n=1}^{\infty}$ does not.

30. Find the limit of the sequences with general term as given:

a. $\dfrac{n^2 + 4n}{n^2 - 5}$ b. $\dfrac{\cos n}{n}$ c. $\dfrac{\sin n^2}{\sqrt{n}}$ d. $\dfrac{n}{n^2 - 3}$

e. $\left(\sqrt{4 - \dfrac{1}{n}} - 2\right)n$ f. $(-1)^n \dfrac{\sqrt{n}}{n + 7}$

31. Find the limit of the sequence in Exercise 21 when $a_0 = 0$ and $a_1 = 3$. You might want to look at Example 0.10.

1.4 SUBSEQUENCES AND MONOTONE SEQUENCES

32. Find a convergent subsequence of the sequence $\left\{(-1)^n\left(1 - \dfrac{1}{n}\right)\right\}_{n=1}^{\infty}$.

33. Suppose x is an accumulation point of $\{a_n : n \in J\}$. Show that there is a subsequence of $\{a_n\}_{n=1}^{\infty}$ that converges to x.

***34.** Let $\{a_n\}_{n=1}^{\infty}$ be a bounded sequence of real numbers. Prove that $\{a_n\}_{n=1}^{\infty}$ has a convergent subsequence. (*Hint:* You may want to use the Bolzano–Weierstrass Theorem.)

***35.** Prove that if $\{a_n\}_{n=1}^{\infty}$ is decreasing and bounded, then $\{a_n\}_{n=1}^{\infty}$ converges.

36. Prove that if $c > 1$, then $\{\sqrt[n]{c}\}_{n=1}^{\infty}$ converges to 1.

***37.** Suppose $\{x_n\}_{n=1}^{\infty}$ converges to x_0 and $\{y_n\}_{n=1}^{\infty}$ converges to x_0. Define a sequence $\{z_n\}_{n=1}^{\infty}$ as follows: $z_{2n} = x_n$ and $z_{2n-1} = y_n$. Prove that $\{z_n\}_{n=1}^{\infty}$ converges to x_0.

38. Show that the sequence defined by $a_1 = \sqrt{6}$ and $a_n = \sqrt{6 + a_{n-1}}$ for $n > 1$ is convergent and find the limit.

***39.** Show that if x is any real number, there is a sequence of rational numbers converging to x.

***40.** Show that if x is any real number, there is a sequence of irrational numbers converging to x.

41. Suppose that $\{a_n\}_{n=1}^{\infty}$ converges to A and that B is an accumulation point of $\{a_n : n \in J\}$. Prove that $A = B$.

MISCELLANEOUS

42. Suppose that $\{a_n\}_{n=1}^{\infty}$ and $\{b_n\}_{n=1}^{\infty}$ are two sequences of positive real numbers. We say that a_n is $O(b_n)$ (read as a_n is "big oh" of b_n) if there is an integer N and a real number M such that for $n \geqslant N$, $a_n \leqslant M \cdot b_n$. Prove that if $\left\{\dfrac{a_n}{b_n}\right\}_{n=1}^{\infty}$ converges to $L \neq 0$, then a_n is $O(b_n)$ and b_n is $O(a_n)$. What can you say if $L = 0$? Illustrate with examples.

PROJECT

The purpose of this project is to prove that the sequence $\left\{\left(1 + \dfrac{1}{n}\right)^n\right\}_{n=1}^{\infty}$ converges. The strategy is to prove that the sequence is increasing and bounded. The limit is e, the base for the natural logarithm. The project in Chapter 5 will shed more light on this function.

We use the binomial theorem to write

$$\left(1 + \frac{1}{n}\right)^n = 1 + 1 + \frac{n(n-1)}{2!}\left(\frac{1}{n}\right)^2 + \cdots + \binom{n}{k}\left(\frac{1}{n}\right)^k + \cdots + \left(\frac{1}{n}\right)^n$$

where $\dbinom{n}{k} = \dfrac{n(n-1)(n-2)\cdot \cdots \cdot (n-k+1)}{k!}$.

1. Prove that for $k \geqslant 2$, $\dbinom{n}{k}\left(\dfrac{1}{n}\right)^k \leqslant \dfrac{1}{k!} \leqslant \dfrac{1}{2^{k-1}}$

2. Prove that $1 + \dfrac{1}{2} + \dfrac{1}{4} + \cdots + \dfrac{1}{2^{n-1}} \leqslant 2$ for all $n \in J$.

3. Prove that the sequence $\left\{\left(1 + \dfrac{1}{n}\right)^n\right\}_{n=1}^{\infty}$ is bounded.

4. Prove that for all $k \geqslant 2$, $\dbinom{n+1}{k}\left(\dfrac{1}{n+1}\right)^k \geqslant \dbinom{n}{k}\left(\dfrac{1}{n}\right)^k$

5. Prove that the sequence $\left\{\left(1 + \dfrac{1}{n}\right)^n\right\}_{n=1}^{\infty}$ is increasing.

6. Items 3 and 5 show that the sequence $\left\{\left(1 + \dfrac{1}{n}\right)^n\right\}_{n=1}^{\infty}$ is bounded and increasing, hence it is convergent; call the limit e. Compute the 100th, 1000th, and 100,000th term of the sequence to approximate e. Compare your results with the number e that your calculator will display.

CHAPTER 2

Limits of Functions

The notion of the existence of a limit of a function at a point underlies the study of calculus. When students fail to understand this idea, their study of calculus becomes a drudgery of juggling formulas. Before considering continuity, differentiation, and integration, one needs a thorough understanding of limits. The material on sequences presented in Chapter 1 will help in achieving this understanding.

First we shall consider an example commonly found in calculus books. Let

$$f(x) = \frac{x^2 - 1}{x - 1}$$

for $x \neq 1$, and let $f(1) = 6$. We find that $f(x) = x + 1$ for $x \neq 1$ and $f(1) = 6$. To the unsophisticated, it may seem objectionable to define a function this way. Since f behaves so well elsewhere, it seems reasonable to define $f(1) = 1 + 1 = 2$. In this chapter we shall attempt to rearrange any prejudices of the reader concerning the behavior of such functions.

2.1 DEFINITION OF THE LIMIT OF A FUNCTION

First of all, let us examine the intuitive idea of a limit of a function that was presented in elementary calculus. Consider a function $f\colon D \to R$ where $D \subset R$ and L is a real number.† If $x_0 \in R$, then f has a limit L at x_0 if, as x approaches x_0, $f(x)$ gets close to L. Of course, the values of x to be considered must belong to the domain of the function. The precise formulation of the idea of getting close no longer poses any problem, since this has already been done in Chapter 1 while dealing with convergent sequences. To summarize, in order to formalize the definition of a limit of a function, it is necessary to give a precise definition of what

†Here and throughout the remainder of this book, R will be the set of all real numbers, and the domain of any functions considered will be subsets of R.

is meant by "$f(x)$ gets close to L as x approaches x_0." To set the stage for this, it is necessary that there be points in D as close to x_0 as one may wish. Finally, the limit should depend only on the behavior of the function near x_0, not on $f(x_0)$; indeed, x_0 need not even belong to D.

DEFINITION Let $f: D \to R$ with x_0 an accumulation point of D. Then f has a limit L at x_0 iff for each $\epsilon > 0$ there is a $\delta > 0$ such that if $0 < |x - x_0| < \delta$ and $x \in D$, then $|f(x) - L| < \epsilon$.

Observe that the definition is similar to that concerning a sequence converging to a real number L, in that it does not specify that if L exists, it is unique. However, this is true and the proof is left as Exercise 5. If f has a limit L at x_0, we write $L = \lim_{x \to x_0} f(x)$.

It is incorrect for students to think that they should memorize all definitions such as the preceding one. That this mistaken impression is widespread among students is evidenced by the fact that many are able to regurgitate the words of the definition, sometimes in the wrong order, without understanding the full meaning. A definition should be carefully examined, dissected, and fully assimilated. The reasons behind it and the full meaning of each part of the statement should be carefully considered. The ideas should become not only a part of one's memory, but a part of one's way of thinking—familiar friends and companions along the journey to enjoyment of mathematics.

First of all, the insistence that x_0 be an accumulation point of D, the domain of f, is necessary in order that f be defined at points near x_0. Now, the positive number ϵ is the desired degree of closeness chosen in advance. To fulfill the definition, one must find $\delta > 0$ (which, of course, will generally depend on ϵ) such that if $x \in (x_0 - \delta, x_0 + \delta)$, $x \in D$, and $x \neq x_0$ (the behavior of f at x_0 should not affect the limit); then $f(x) \in (L - \epsilon, L + \epsilon)$. This last statement is equivalent to insisting that the graph of f for $x \in D \cap (x_0 - \delta, x_0 + \delta)$, $x \neq x_0$, lies in the strip $\{(x, y) : L - \epsilon < y < L + \epsilon\}$. See Figure 2.1. The hole in the graph is to emphasize that the value of f at x_0 is not considered when determining whether L is the limit of f at x_0.

■ **Example 2.1** Let us now reconsider the example mentioned at the beginning of the chapter. Let $D = R$, and define $f: D \to R$ by

$$f(x) = \frac{x^2 - 1}{x - 1}$$

for $x \neq 1$ and $f(1) = 6$. For $x \neq 1$, $f(x) = \dfrac{x^2 - 1}{x - 1} = x + 1$; hence f is a linear function, and the graph of f is a line with slope 1, except for $x = 1$. As mentioned above, the number $f(1)$ has nothing to do with the existence of a limit at $x = 1$. Thus, one is led to believe (correctly, by the geometry of the plane) that as x approaches 1, $f(x)$ approaches the value necessary to fill in the gap in the line—namely, $1 + 1 = 2$. Let us prove that f has a limit $L = 2$ at $x = 1$.

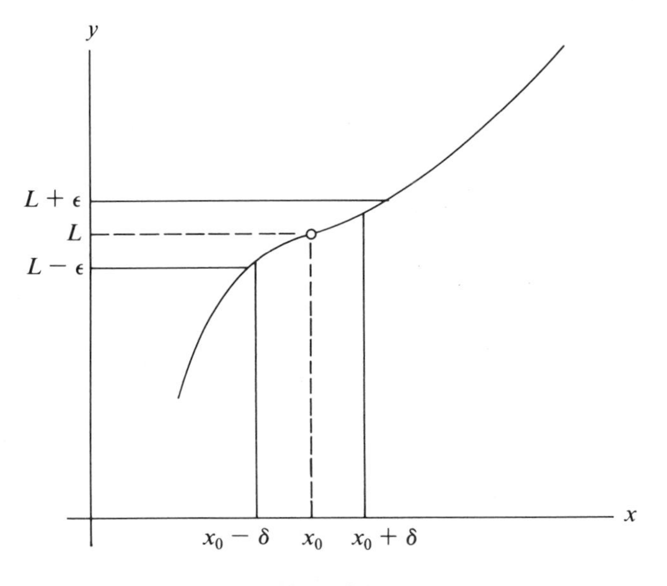

Figure 2.1

Choose $\epsilon > 0$. Consider the geometric interpretation of the idea of a limit. It is necessary to choose a neighborhood of 1 such that for x in this neighborhood with $x \neq 1$, the corresponding points on the graph of f lie in the strip $\{(x, y) : 2 - \epsilon < y < 2 + \epsilon\}$. Now ignoring the point $x = 1$, the graph of f is a straight line of slope 1. Thus, one is tempted to try $\epsilon = \delta$ to obtain the neighborhood $(1 - \delta, 1 + \delta)$ of $x = 1$. This result could be obtained by direct computation, but it is worthwhile to take this opportunity to emphasize the geometric aspect of the definition. If $0 < |x - 1| < \delta = \epsilon$, then

$$|f(x) - 2| = \left|\frac{x^2 - 1}{x - 1} - 2\right| = |(x + 1) - 2| = |x - 1| < \delta = \epsilon,$$

and the conjecture is shown to be correct. See Figure 2.2. ■

■ **Example 2.2** Define $f : [0, 1] \to R$ by $f(x) = 0$ for x rational and $f(x) = 1$ for x irrational. Every interval in $[0, 1]$ contains both rational and irrational points; hence, in every interval, f assumes the values 0 and 1. Now if f were to have a limit L at some point $x_0 \in [0, 1]$, then all values of f at points near x_0 must lie close to L. Let $\epsilon = \frac{1}{4}$. Then there must be $\delta > 0$ such that for $0 < |x - x_0| < \delta$ and $x \in [0, 1]$, we have $|f(x) - L| < \frac{1}{4}$. Choose r rational and p irrational such that $0 < |r - x_0| < \delta$, $r \in [0, 1]$, and $0 < |p - x_0| < \delta$, $p \in [0, 1]$. Then $f(r) = 0$ and $f(p) = 1$, so

$$\begin{aligned} 1 = |f(p) - f(r)| &\leq |f(p) - L + L - f(r)| \\ &\leq |f(p) - L| + |L - f(r)| < \frac{1}{4} + \frac{1}{4} = \frac{1}{2}. \end{aligned}$$

This is obviously false, so f does not have a limit at x_0. ■

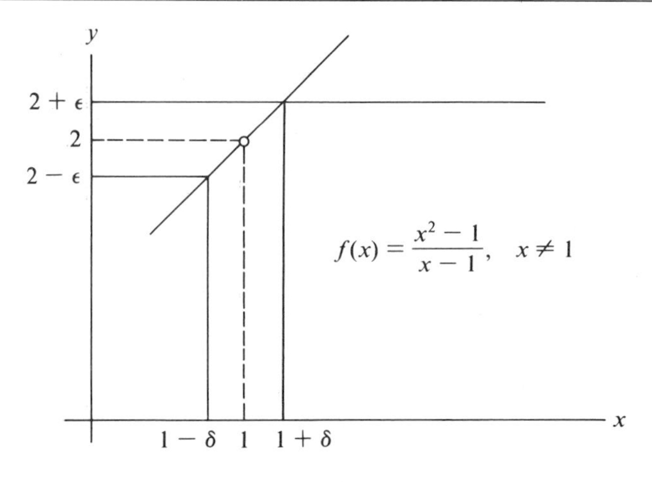

Figure 2.2

■ **Example 2.3** Let us now consider a more sophisticated example. Define f: $[0, 1] \to R$ by $f(x) = 0$ if x is irrational; and if x is rational, set $f(x) = \frac{1}{q}$ where $x = \frac{p}{q}$ with p and q nonnegative integers that are relatively prime. Thus, we have $f(0) = 1$, $f(\frac{1}{4}) = \frac{1}{4}$, $f(\frac{2}{3}) = \frac{1}{3}$, $f(\frac{\sqrt{2}}{2}) = 0$, $f(\frac{6}{8}) = \frac{1}{4}$, etc. We shall seek to determine those points at which f has a limit and those at which f does not have a limit.

Suppose $x_0 \in [0, 1]$, and let us examine the behavior of f near x_0. Since there are irrational points in every neighborhood of x_0 that are distinct from x_0, it is clear that f takes the value 0 in every neighborhood of x_0 infinitely often. Thus, if there is to be a limit of f at x_0, it is clear that the limit must be zero. It remains to decide if it is reasonable to expect that f has zero as a limit at x_0. For $x \in [0, 1]$, $f(x)$ is small (close to zero) only if x is irrational or if $x = \frac{p}{q}$ and p and q are relatively prime with q large. However, for a fixed positive integer q, there are but a finite number of points in $[0, 1]$ of the form $\frac{p}{q}$; in fact, for a fixed positive integer q_0, there are only a finite number of points in $[0, 1]$ of the form $\frac{p}{q}$, where $q \leqslant q_0$ with p and q positive integers. Thus, $f(x) \geqslant \frac{1}{q_0}$ at only a finite number of points. Now, in view of these observations, the reader should suspect that indeed f has a limit at x and the limit is zero. Let us now give a proof of this fact.

Choose $\epsilon > 0$. There is a positive integer q_0 such that $\frac{1}{q_0} < \epsilon$. There are at most a finite number of rational points in $[0, 1]$ of the form $\frac{p}{q}$ where p and q are positive integers with $q < q_0$, say $r_1, \ldots, r_n$. We may assume that x_0 is deleted from this list if it should happen to be of this form. Now, to guarantee that $f(x)$ is small, it is sufficient to avoid these points. Thus, let

$$\delta = \min \{|x_0 - r_i| : i = 1, \ldots, n\},$$

and observe that $\delta > 0$. Now if $0 < |x - x_0| < \delta$ and $x \in [0, 1]$, then x is either irrational, in which case $f(x) = 0$, or $x = \frac{p}{q}$ where p and q are relatively prime with $q > q_0$, in which case $f(x) = \frac{1}{q}$. In either case,

$$|f(x) - 0| = |f(x)| \leqslant \frac{1}{q} < \frac{1}{q_0} < \epsilon.$$

Thus, f has a limit at each $x \in [0, 1]$ and that limit is zero.

■ **Example 2.4** Define f: $(0, 1) \to R$ by $f(x) = \sin \frac{1}{x}$, and consider the behavior of f at $x = 0$. Although zero is not in the domain of f, it is an accumulation point of the domain of f; hence, it is reasonable to inquire whether f has a limit at zero. In this example, it is instructive to look at the graph of this function in Figure 2.3 to gain insight into its behavior as x approaches zero. Since $f\left(\frac{1}{n\pi}\right) = 0$ for each positive integer n and $f\left(\frac{2}{n\pi}\right)$ is $+1$ or -1 for n an odd positive integer, we see that the function oscillates more and more wildly as x becomes close to zero. Consequently, it is reasonable to conjecture that f does not have a limit at zero. Let L be any real number and choose $\epsilon = \frac{1}{3}$. To show that L is not a limit of f at zero, it suffices to show that for any $\delta > 0$, there is $x \in (0, 1)$ such that $0 < |x - 0| < \delta$ and $|f(x) - L| \geqslant \epsilon$. By the choice of ϵ, either $1 \not\in (L - \epsilon, L + \epsilon)$ or $0 \not\in (L - \epsilon, L + \epsilon)$. Suppose the latter is the case. Then, given $\delta > 0$, there is a positive integer n such that $\frac{1}{n\pi} < \delta$, and, as observed above, $f\left(\frac{1}{n\pi}\right) = 0 \not\in (L - \epsilon, L + \epsilon)$; hence, $\left|f\left(\frac{1}{n\pi}\right) - L\right| \geqslant \epsilon$. The case for $1 \not\in (L - \epsilon, L + \epsilon)$ may be handled similarly. Thus, f does not have a limit at zero.

A word of philosophy seems appropriate at this point. The emphasis on rigor, in both definitions and proofs, is intentional, since this is the only way mathematics can be communicated intelligently. However, one's feelings and intuition about these matters should not be discarded or considered unimportant. On the contrary, intuition is very important and must be cultivated, but it must also

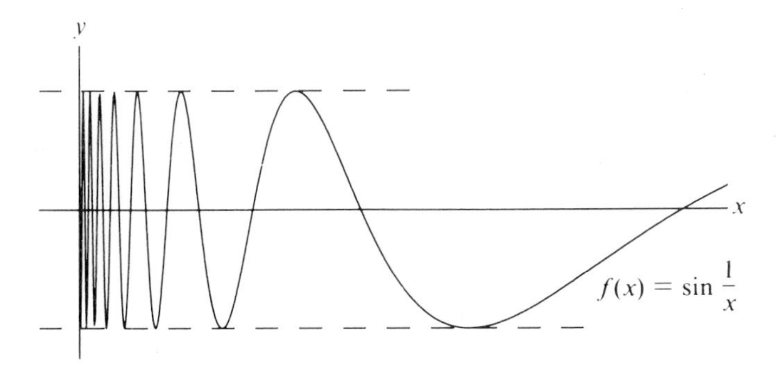

Figure 2.3

be accompanied by the realization that we must provide rigorous proofs for those facts that our intuition tells us are true. Finally, the intuition must constantly be readjusted, because we are occasionally led astray by our feelings. In the preceding example, the oscillatory nature of the function, as evidenced by a sketch of the graph, led our intuition to tell us that the function does not have a limit at zero. The behavior of the function $f(x) = 0$ for x irrational and $f(\frac{p}{q}) = \frac{1}{q}$ for p and q nonnegative integers that are relatively prime is probably an example that will elude the intuition of the novice. However, the facts are now before us and should be considered as a new experience leading to greater mathematical maturity. The following example should be a familiar friend, and all we shall do is fill in the necessary rigor.

■ **Example 2.5** Define f: $(0, 1) \to R$ by $f(x) = \dfrac{1}{x}$. Now zero is an accumulation point of $(0, 1)$; hence, we may inquire as to the existence of the limit of f at zero. Again, from a sketch of a part of the graph of this function, it is seen that the function "blows up" at zero. This is not the sort of behavior one expects a function to have if it is to have a limit at zero, so let us prove that f does not have a limit at zero. Let L be any real number, and choose $\epsilon > 0$ such that $L + \epsilon > 0$. Now, if $0 < x < \dfrac{1}{L + \epsilon}$, then $L + \epsilon < \dfrac{1}{x} = f(x)$; hence, $|f(x) - L| > \epsilon$. Thus, it is impossible to find a $\delta > 0$ to fulfill the requirements of the definition; that is, L is not a limit of f at zero. Since L is any real number, f does not have a limit at zero. See Figure 2.4. ■

The examples thus far considered could be handled more easily with some machinery that can be built up for this purpose. However, the process of considering them directly in terms of the definition should help in the understanding of the concept of the limit of a function.

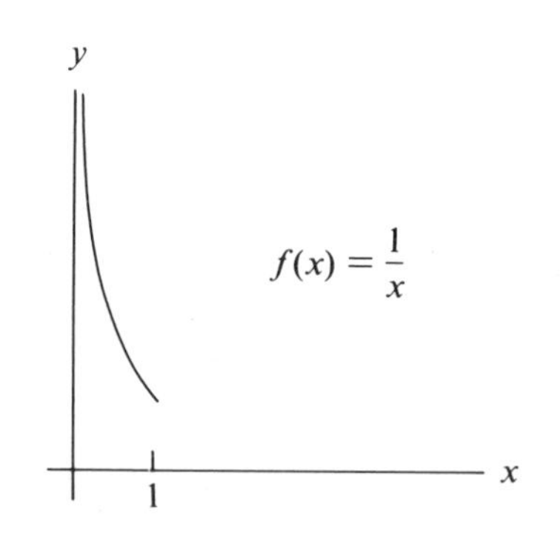

Figure 2.4

2.2 LIMITS OF FUNCTIONS AND SEQUENCES

As mentioned in Chapter 1, a very strong relationship exists between limits of sequences and limits of functions. Now is clearly the time to seek out these relationships and exploit them to the fullest. Suppose $f\colon D \to R$ with x_0 an accumulation point of D, and suppose f has a limit L at x_0. Consider a sequence $\{x_n\}_{n=1}^{\infty}$ converging to x_0 with $x_n \in D$, $x_0 \neq x_n$, for each positive integer n. Since f has a limit L at x_0, as the terms of the sequence get close to x_0, the corresponding values of f must get close to L; in fact, we must have $\{f(x_n)\}_{n=1}^{\infty}$ converging to L. Fortunately, the converse of this theorem is also true. We summarize these results in the following theorem.

2.1 THEOREM Let $f\colon D \to R$ with x_0 an accumulation point of D. Then f has a limit at x_0 iff for each sequence $\{x_n\}_{n=1}^{\infty}$ converging to x_0 with $x_n \in D$ and $x_n \neq x_0$ for all n, the sequence $\{f(x_n)\}_{n=1}^{\infty}$ converges.

Before proving this theorem, reflect on one aspect of its content. Consider two sequences $\{x_n\}_{n=1}^{\infty}$ and $\{y_n\}_{n=1}^{\infty}$ such that $x_n, y_n \in D$, $x_n \neq x_0$, $y_n \neq x_0$ for $n = 1, 2, \ldots$, and such that both $\{x_n\}_{n=1}^{\infty}$ and $\{y_n\}_{n=1}^{\infty}$ converge to x_0. If we assume that the latter condition of the theorem holds, then $\{f(x_n)\}_{n=1}^{\infty}$ and $\{f(y_n)\}_{n=1}^{\infty}$ both converge to, say, L_1 and L_2, respectively. (Note that the condition does not assert that $L_1 = L_2$.) Form a new sequence $\{z_n\}_{n=1}^{\infty}$ where $z_{2n} = x_n$ and $z_{2n-1} = y_n$. This sequence consists of members of D distinct from x_0, and it converges to x_0; hence $\{f(z_n)\}_{n=1}^{\infty}$ converges. In particular, $\{f(x_n)\}_{n=1}^{\infty}$ and $\{f(y_n)\}_{n=1}^{\infty}$ are subsequences of the convergent sequence $\{f(z_n)\}_{n=1}^{\infty}$ and hence have the same limit; that is, $L_1 = L_2$. Now, if for every sequence $\{x_n\}_{n=1}^{\infty}$ of members of D distinct from x_0 and converging to x_0, the sequence $\{f(x_n)\}_{n=1}^{\infty}$ converges, then all such sequences have a common limit by use of the above observations. This limit should be the limit of the function, a fact that remains to be seen.

Proof. Suppose f has a limit L at x_0. Let $\{x_n\}_{n=1}^{\infty}$ be a sequence of members of D distinct from x_0 but converging to x_0, and consider the sequence $\{f(x_n)\}_{n=1}^{\infty}$. Choose $\epsilon > 0$. There is $\delta > 0$ such that if $0 < |x - x_0| < \delta$ with $x \in D$, then $|f(x) - L| < \epsilon$. Since $\{x_n\}_{n=1}^{\infty}$ converges to x_0, there is N such that for $n \geqslant N$, $|x_n - x_0| < \delta$. Now for $n \geqslant N$, $0 < |x_n - x_0| < \delta$ and $x_n \in D$; hence, $|f(x_n) - L| < \epsilon$. Thus, $\{f(x_n)\}_{n=1}^{\infty}$ converges; indeed, it converges to L.

Suppose now that the latter condition is satisfied; and by the remarks following the theorem, all the sequences $\{f(x_n)\}_{n=1}^{\infty}$ have a common limit, called (with great originality) L. Suppose that L is not a limit of f at x_0. (We do not assume anything concerning the existence of the limit of f at x_0; we assume only that L is not a limit of f at x_0.) Thus, there is $\epsilon > 0$ such that for every $\delta > 0$, there is $x \in D$, with $0 < |x - x_0| < \delta$ and such that

$|f(x) - L| \geq \epsilon$. In particular, for each positive integer n, there is $x_n \in D$ with $0 < |x_n - x_0| < \frac{1}{n}$ such that $|f(x_n) - L| \geq \epsilon$. The sequence $\{x_n\}_{n=1}^{\infty}$ converges to x_0 and is a sequence of members of D distinct from x_0; hence, $\{f(x_n)\}_{n=1}^{\infty}$ converges to L, contrary to the fact that $|f(x_n) - L| \geq \epsilon > 0$ for all n. Thus, L must be the limit of f at x_0.

To summarize, f has a limit of L at x_0 iff, for each sequence $\{x_n\}_{n=1}^{\infty}$ of members of D distinct from x_0 converging to x_0, the sequence $\{f(x_n)\}_{n=1}^{\infty}$ converges. If the second condition is satisfied, then all the given sequences $\{f(x_n)\}_{n=1}^{\infty}$ have a common limit, which is L, the limit of the function f. An example should serve to reveal the usefulness of this equivalence and the related observation that follows it. Let f: $(0, 1) \rightarrow R$ satisfy the following condition: there is $K > 0$ such that, for all $x, y \in (0, 1)$, $|f(x) - f(y)| \leq K|x - y|$. Thus, in particular, whenever $\{x_n\}_{n=1}^{\infty}$ is a Cauchy sequence in $(0, 1)$, then so is $\{f(x_n)\}_{n=1}^{\infty}$; hence, if $\{x_n\}_{n=1}^{\infty}$ is a sequence in $(0, 1)$ converging to zero, the sequence $\{f(x_n)\}_{n=1}^{\infty}$ is convergent. By the theorem, f has a limit at zero; and that may be computed by determining the limit of $\{f(x_n)\}_{n=1}^{\infty}$ for any sequence $\{x_n\}_{n=1}^{\infty}$ in $(0, 1)$ converging to zero. A judicious choice of the sequence $\{x_n\}_{n=1}^{\infty}$ may make the task less tedious. See Exercise 10 for an application of this idea.

The urge to reconsider the preceding remarks and to generalize the result is irresistible. The conclusion stems from the fact that if $\{x_n\}_{n=1}^{\infty}$ is Cauchy, then $\{f(x_n)\}_{n=1}^{\infty}$ is Cauchy. The proof of the following theorem is immediate.

2.2 THEOREM Let f: $D \rightarrow R$, and suppose x_0 is an accumulation point of D. If for each sequence $\{x_n\}_{n=1}^{\infty}$ converging to x_0 with $x_n \in D \setminus \{x_0\}$ for each n, the sequence $\{f(x_n)\}_{n=1}^{\infty}$ is Cauchy, then f has a limit at x_0.

Let us now sift through our newly gained knowledge of sequences and discover theorems concerning limits of functions. Many of these theorems may be proved by exploiting Theorem 2.1, but in some cases we will give a direct proof in order to introduce the reader to the act of finding δ's.

2.3 THEOREM Let f: $D \rightarrow R$, with x_0 an accumulation point of D. If f has a limit at x_0, then there is a neighborhood Q of x_0 and a real number M such that for all $x \in Q \cap D$, $|f(x)| \leq M$.

Proof. Let $\epsilon = 1$, and let L be the limit of f at x_0. Then there is $\delta > 0$ such that for $0 < |x - x_0| < \delta$ and $x \in D$, we have $|f(x) - L| < \epsilon = 1$. If $x_0 \in D$, define

$$M = \max\{|L - 1|, |L + 1|, |f(x_0)|\};$$

otherwise, define

$$M = \max\{|L - 1|, |L + 1|\},$$

and let $Q = (x_0 - \delta, x_0 + \delta)$. In either case, if $x \in Q \cap D$, then $|f(x)| \leq M$.

In less formal language, if f has a limit at x_0, then f is bounded near x_0. In particular, this gives another verification that $f(x) = \frac{1}{x}$ does not have a limit at zero since f is unbounded in every neighborhood of zero.

■ **Example 2.6** For each real number x, let $[x]$ denote the largest integer that is less than or equal to x. For example, $[\pi] = 3$, $[\frac{3}{2}] = 1$, and $[2] = 2$. For $n \leq x < n + 1$ with n an integer, $[x] = n$, so the graph of the function looks like a set of stairs with a "jump" of one unit at each integer. See Figure 2.5 for a sketch of this function. If we have been successful in developing the geometric concept of the limit of a function, it should be apparent that $f(x) = [x]$ has a limit at x_0 iff x_0 is not an integer. Suppose x_0 is not an integer and let δ be the distance from x_0 to the nearest integer. Now $\delta > 0$, and if $0 < |x - x_0| < \delta$, then $[x] = [x_0]$; hence $|f(x) - [x_0]| = 0 < \epsilon$ for all $\epsilon > 0$. Thus, f has a limit at x_0 if x_0 is not an integer. If x_0 is an integer, consider the sequence $\left\{x_0 + (-1)^n \frac{1}{n}\right\}_{n=1}^{\infty}$. Now, for n odd, $n > 1$, $x_0 + (-1)^n \frac{1}{n} < x_0$, and thus

$$f\left(x_0 + (-1)^n \frac{1}{n}\right) = x_0 - 1;$$

and for n even, $x_0 + (-1)^n \frac{1}{n} > x_0$, and thus

$$f\left(x_0 + (-1)^n \frac{1}{n}\right) = x_0.$$

Thus, $\left\{f\left(x_0 + (-1)^n \frac{1}{n}\right)\right\}_{n=1}^{\infty}$ does not converge, and so f does not have a limit at x_0 if x_0 is an integer.

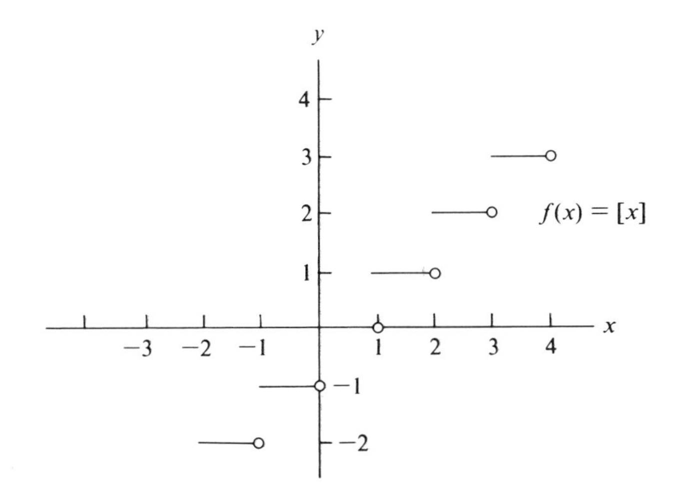

Figure 2.5

Consider the examples $f(x) = [x]$, $g(x) = \sin \frac{1}{x}$, and $h(x) = \frac{1}{x}$. Each failed to have a limit at zero for basically different reasons. The function $f(x) = [x]$ "jumped" at zero, taking the value -1 to the left of zero and the value 0 to the right of zero; g oscillated too badly at zero to have a limit there; h was unbounded near zero and so could not have a limit there. In a sense, these three examples describe the sort of behavior one may suspect of a function that does not have a limit at a point. We shall comment more on this in Section 2.4.

2.3 ALGEBRA OF LIMITS

Following the pattern set in Chapter 1, we shall now determine the relationships between limits of functions and the algebraic operations. If $f: D \to R$ and $g: D \to R$, define $f + g: D \to R$ by $(f + g)(x) = f(x) + g(x)$ for all $x \in D$. In a similar fashion, define $fg: D \to R$ by $(fg)(x) = f(x)g(x)$. If $g(x) \neq 0$ for $x \in D$, we may define $\frac{f}{g}: D \to R$ by $\left(\frac{f}{g}\right)(x) = \frac{f(x)}{g(x)}$ for all $x \in D$. Now read Theorems 1.8, 1.9, and 1.11 and attempt to formulate corresponding theorems for limits of functions. In each case, the proof can be obtained by use of Theorem 2.1 and the appropriate result from Chapter 1.

2.4 THEOREM Suppose $f, g: D \to R$ with x_0 an accumulation point of D, and further suppose that f and g have limits at x_0. Then,

1. $f + g$ has a limit at x_0 and
$$\lim_{x \to x_0} (f + g)(x) = \lim_{x \to x_0} f(x) + \lim_{x \to x_0} g(x);$$
2. fg has a limit at x_0 and
$$\lim_{x \to x_0} (fg)(x) = [\lim_{x \to x_0} f(x)][\lim_{x \to x_0} g(x)];$$
3. If $g(x) \neq 0$ for all $x \in D$ and $\lim_{x \to x_0} g(x) \neq 0$, then $\frac{f}{g}$ has a limit at x_0, and
$$\lim_{x \to x_0} \left(\frac{f}{g}\right)(x) = \frac{\lim_{x \to x_0} f(x)}{\lim_{x \to x_0} g(x)}.$$

The proofs of (1) and (3) will be accomplished by use of sequences, whereas the proof of (2) will be direct.

Proof. 1. Let $\{x_n\}_{n=1}^{\infty}$ be any sequence of points in D converging to x_0 with $x_n \neq x_0$ for all n. It suffices to show that $\{(f + g)(x_n)\}_{n=1}^{\infty}$ converges to
$$\lim_{x \to x_0} f(x) + \lim_{x \to x_0} g(x).$$

By assumption, f and g have limits at x_0; hence, $\{f(x_n)\}_{n=1}^{\infty}$ converges to $\lim_{x \to x_0} f(x)$ and $\{g(x_n)\}_{n=1}^{\infty}$ converges to $\lim_{x \to x_0} g(x)$. By Theorem 1.8,

$$\{(f + g)(x_n)\}_{n=1}^{\infty} = \{f(x_n) + g(x_n)\}_{n=1}^{\infty}$$

converges to $\lim_{x \to x_0} f(x) + \lim_{x \to x_0} g(x)$. This concludes the proof of (1).

2. Given $\epsilon > 0$, we must find an appropriate $\delta > 0$. The reader is referred to the discussion preceding Theorem 1.9 before beginning the proof. As the plan for finding the δ unfolds, observe the parallel between this search and that for N in Theorem 1.9.

Choose $\epsilon > 0$. Let

$$A = \lim_{x \to x_0} f(x) \quad \text{and} \quad B = \lim_{x \to x_0} g(x).$$

We wish to find $\delta > 0$ such that if $0 < |x - x_0| < \delta$ and $x \in D$, then $|f(x)g(x) - AB| < \epsilon$. By Theorem 2.3, there is $\delta_1 > 0$ and a real number $M > 0$ such that for $0 < |x - x_0| < \delta_1$, $x \in D$, we have $|f(x)| \leq M$. Let

$$\epsilon' = \frac{\epsilon}{|B| + M} > 0.$$

There is $\delta_2 > 0$ such that if $0 < |x - x_0| < \delta_2$, $x \in D$, then $|f(x) - A| < \epsilon'$; and there is $\delta_3 > 0$ such that for $0 < |x - x_0| < \delta_3$, $x \in D$, we have $|g(x) - B| < \epsilon'$. Let

$$\delta = \min \{\delta_1, \delta_2, \delta_3\}.$$

Now if $0 < |x - x_0| < \delta$, $x \in D$, then,

$$\begin{aligned} |(fg)(x) - AB| &= |f(x)g(x) - AB| \leq |f(x)g(x) - f(x)B| + |f(x)B - AB| \\ &= |f(x)|\,|g(x) - B| + |B|\,|f(x) - A| < M\epsilon' + |B|\epsilon' \\ &= \epsilon. \end{aligned}$$

3. Suppose $\{x_n\}_{n=1}^{\infty}$ is a sequence of members of D, distinct from x_0, converging to x_0. Now $\{f(x_n)\}_{n=1}^{\infty}$ converges to $\lim_{x \to x_0} f(x)$ and $\{g(x_n)\}_{n=1}^{\infty}$ converges to $\lim_{x \to x_0} g(x)$. Since $g(x) \neq 0$ for all $x \in D$, $g(x_n) \neq 0$ for all n. By assumption, $\lim_{x \to x_0} g(x) \neq 0$. Hence

$$\left\{\left(\frac{f}{g}\right)(x_n)\right\}_{n=1}^{\infty} = \left\{\frac{f(x_n)}{g(x_n)}\right\}_{n=1}^{\infty}$$

converges to

$$\frac{\lim_{x \to x_0} f(x)}{\lim_{x \to x_0} g(x)}.$$

By Theorem 2.1, $\frac{f}{g}$ has the indicated limit at x_0.

The adventuresome reader may wish to attempt a direct proof of (3). A lemma similar to Lemma 1.10 is in order, and then the proof may be patterned after that of Theorem 1.11. See Exercise 21. Good luck!

2.5 THEOREM Suppose $f: D \to R$ and $g: D \to R$, x_0 is an accumulation point of D, and f and g have limits at x_0. If $f(x) \leq g(x)$ for all $x \in D$, then

$$\lim_{x \to x_0} f(x) \leq \lim_{x \to x_0} g(x).$$

Proof. The proof is left as Exercise 20.

Let us now look at an example suggested by the sequence $\left\{\frac{1}{n} \sin \frac{n\pi}{4}\right\}_{n=1}^{\infty}$. Consider the function $f: (0, 1) \to R$, defined by $f(x) = x \sin \frac{1}{x}$. It has been shown that $\sin \frac{1}{x}$ fails to have a limit at zero; so we may not use Theorem 2.4 (2). However $\sin \frac{1}{x}$ is bounded above by 1 and below by -1. Now it is clear that

$$|f(x)| = \left|x \sin \frac{1}{x}\right| \leq |x|;$$

hence, f has a limit at zero; in fact, $\lim_{x \to x_0} f(x) = 0$. (We merely choose $\delta = \epsilon$.) Now, a theorem reminiscent of Theorem 1.13 is in order.

2.6 THEOREM Let $f: D \to R$ and $g: D \to R$ and x_0 be an accumulation point of D. If f is bounded in a neighborhood of x_0 and g has limit zero at x_0, then fg has a limit at x_0 and $\lim_{x \to x_0} fg(x) = 0$.

Proof. Choose $\epsilon > 0$. There is a $\delta_1 > 0$ and $M > 0$ such that if $x \in D$ and $|x - x_0| < \delta_1$, then $|f(x)| \leq M$. Let $\epsilon' = \frac{\epsilon}{M}$. There is $\delta_2 > 0$ such that if $x \in D$ and $0 < |x - x_0| < \delta_2$, then

$$|g(x) - \lim_{x \to x_0} g(x)| = |g(x)| < \epsilon'.$$

Choose $\delta = \min\{\delta_1, \delta_2\}$. Then if $0 < |x - x_0| < \delta$ with $x \in D$, a computation yields

$$|(fg)(x)| = |f(x)g(x)| = |f(x)|\,|g(x)| \leq M\epsilon' = \epsilon.$$

Hence, fg has a limit at x_0, and $\lim_{x \to x_0} (fg)(x) = 0$.

The proper use of the preceding theorems will allow us to handle a rather large class of functions. Suppose $f: R \to R$ is defined by $f(x) = x$ for each

$x \in R$. If $x_0 \in R$, then for each sequence $\{x_n\}_{n=1}^{\infty}$ converging to x_0 with $x_n \in R \setminus \{x_0\}$,

$$\{f(x_n)\}_{n=1}^{\infty} = \{x_n\}_{n=1}^{\infty}$$

is convergent to x_0, hence f has a limit at x_0; in particular,

$$\lim_{x \to x_0} f(x) = x_0.$$

Now this is not intended to startle you; it is included for completeness. Choose $k \in R$ and consider the function $g\colon R \to R$ where $g(x) = k$ for all $x \in R$. It is clear that g has a limit at x_0 for each $x_0 \in R$ and $\lim_{x \to x_0} g(x) = k$. These last two statements are usually expressed as

$$\lim_{x \to x_0} x = x_0 \qquad \text{and} \quad \lim_{x \to x_0} k = k.$$

Now by use of induction and Theorem 2.4 (2), we see that the function $f(x) = x^n$ has a limit at x_0 for each $x_0 \in R$, and $\lim_{x \to x_0} x^n = x_0^n$. By using this result, induction, and parts (1) and (2) of Theorem 2.4, we see that for each polynomial

$$p(x) = a_0 + a_1 x + \cdots + a_n x^n,$$

p has a limit at x_0 for each $x_0 \in R$ and $\lim_{x \to x_0} p(x) = p(x_0)$.

If p and q are polynomials and $\{r_1, \ldots, r_n\}$ are the real roots of the equation $q(x) = 0$, then $f\colon R \setminus \{r_1, \ldots, r_n\} \to R$ defined by

$$f(x) = \frac{p(x)}{q(x)}$$

has a limit at every point x_0 of $R \setminus \{r_1, \ldots, r_n\}$ and

$$\lim_{x \to x_0} f(x) = f(x_0) = \frac{p(x_0)}{q(x_0)}.$$

We are safe in using part (3) of Theorem 2.4 in this case since p and q have limits at every point, and $r_1, \ldots, r_n$ are the only points where $\lim_{x \to x_0} q(x) = 0$. The question of the existence of a limit at the points r_i, $i = 1, \ldots, n$, must be handled in a different fashion. We shall postpone the general case until we discuss continuity in Chapter 3.

Recall from Exercise 26 in Chapter 1 that if $\{a_n\}_{n=1}^{\infty}$ is a sequence of nonnegative real numbers converging to a, then $\{\sqrt{a_n}\}_{n=1}^{\infty}$ converges to $\sqrt{a}$. Applying this result to this chapter, it becomes clear that if $f\colon D \to R$ with $f(x) \geqslant 0$ for all $x \in D$ and if f has a limit at x_0, then the function $g(x) = \sqrt{f(x)}$ has a limit at x_0, and

$$\lim_{x \to x_0} \sqrt{f(x)} = \sqrt{\lim_{x \to x_0} f(x)}.$$

■ **Example 2.7** Consider the function h: $(0, 1) \to R$ defined by $h(x) = \dfrac{\sqrt{4 + x} - 2}{x}$. The function h is represented as the quotient of two functions, but, unfortunately, both have limit 0 at zero, so Theorem 2.4 is useless at first glance. Let us attempt to follow a pattern used for sequences by trying to write h in some other form. Now

$$h(x) = \frac{(\sqrt{4 + x} - 2)}{x} = \frac{(\sqrt{4 + x} - 2)(\sqrt{4 + x} + 2)}{x(\sqrt{4 + x} + 2)}$$
$$= \frac{4 + x - 4}{x(\sqrt{4 + x} + 2)} = \frac{1}{\sqrt{4 + x} + 2}.$$

The numerator is the constant function whose value is 1 at each point. By our previous remarks, the denominator is a function that has a limit at zero, namely

$$\lim_{x \to 0} (\sqrt{4 + x} + 2) = \sqrt{4} + 2 = 4.$$

Thus, by Theorem 2.4, h has a limit at zero and

$$\lim_{x \to 0} h(x) = \lim_{x \to 0} \frac{\sqrt{4 + x} - 2}{x} = \frac{1}{4}.$$ ■

Much more about limits will come in later chapters; we end this chapter with a rather interesting theorem.

Previously, we considered three functions that failed to have a limit at a point—one that has a jump, one that is oscillatory, and one that is unbounded. Let us take the first case and seek to determine what behavior can be predicted. Recall that the function $f(x) = 1$ for x irrational and $f(x) = 0$ for x rational had a limit at no point of its domain.

2.4 LIMITS OF MONOTONE FUNCTIONS

DEFINITION Let f: $D \to R$. The function f is said to be *increasing* (*decreasing*) iff, for all $x, y \in D$ with $x \leq y$,

$$f(x) \leq f(y) \quad (f(x) \geq f(y)).$$

If f is either increasing or decreasing, then f is said to be *monotone*.

The function $f(x) = [x]$ is an increasing function, and the only points where f fails to have a limit are the integer points, a countable set. Consider now f: $[\alpha, \beta] \to R$ where f is increasing. Now for all $x \in [\alpha, \beta]$, $f(\alpha) \leq f(x) \leq f(\beta)$, hence f is bounded; and by the monotonicity f cannot be oscillatory. If we attempt to sketch the graph of some functions increasing on $[0, 1]$ that fail to have a limit at some point, after a few such attempts it becomes apparent that the only way for such a function to fail to have a limit at x_0 is to jump at x_0 in a manner similar to the behavior of $f(x) = [x]$ at each integer. Let us get a firmer grip on this idea. For $\alpha < x < \beta$, define $U(x) = \inf \{f(y) : x < y\}$; and for $\alpha < x < \beta$, define $L(x) = \sup \{f(y) : y < x\}$. (Since $f(\alpha) \leq f(x) \leq f(\beta)$ for all $x \in [\alpha, \beta]$, $L(x)$ and $U(x)$ are defined.) $U(x) - L(x)$ measures the jump of f at x; in fact, we shall see in Lemma 2.7 that if $x_0 \in (\alpha, \beta)$, then f has a limit at x_0 iff $U(x_0) - L(x_0) = 0$. Assuming this to be the case, let

$$J_n = \left\{x \in (\alpha, \beta) : U(x) - L(x) > \frac{1}{n}\right\}.$$

Suppose $\{x_1, \ldots, x_k\} \subset J_n$ with $x_1 < x_2 < \cdots < x_k$. Now if f has a jump greater than $\frac{1}{n}$ at $x_1, \ldots, x_k$, then it seems reasonable to suspect that k can't be too large; in fact k should be less than or equal to $n[f(\beta) - f(\alpha)]$. Thus, J_n should be finite, and the set of points where f fails to have a limit should be $\bigcup_{n=1}^{\infty} J_n$, a countable union of finite sets, hence a set that is countable. This is the content of the next theorem; we have outlined the proof in a vague way and must now fill in the details. First we prove a lemma.

2.7 LEMMA Let $f: [\alpha, \beta] \to R$ be increasing. Let

$$U(x) = \inf \{f(y) : x < y\} \quad \text{and} \quad L(x) = \sup \{f(y) : y < x\}$$

for $x \in (\alpha, \beta)$. Then f has a limit at $x_0 \in (\alpha, \beta)$ iff $U(x_0) = L(x_0)$, and in this case

$$\lim_{x \to x_0} f(x) = f(x_0) = U(x_0) = L(x_0).$$

Proof. Suppose f has a limit at $x_0 \in (\alpha, \beta)$, called A. Choose $\epsilon > 0$. There is $\delta > 0$ such that if $0 < |x - x_0| < \delta$ with $x \in [\alpha, \beta]$, then $|f(x) - A| < \epsilon$. Since $x_0 \in (\alpha, \beta)$, there are $x, y \in [\alpha, \beta]$ such that $x_0 - \delta < x < x_0 < y < x_0 + \delta$; so, by the definition of L and U and the fact that f is increasing,

$$A - \epsilon < f(x) \leq L(x_0) \leq f(x_0) \leq U(x_0) \leq f(y) < A + \epsilon.$$

Therefore, $U(x_0) - L(x_0) < 2\epsilon$ for each $\epsilon > 0$; hence $U(x_0) = L(x_0)$; hence $L(x_0) \leq f(x_0) \leq U(x_0)$,

$$L(x_0) = f(x_0) = U(x_0).$$

Moreover, $A - \epsilon < U(x_0) < A + \epsilon$ for all $\epsilon > 0$; hence $U(x_0) = A$. Thus

$$\lim_{x \to x_0} f(x) = U(x_0) = L(x_0) = f(x_0).$$

Suppose now that $U(x_0) = L(x_0)$. As observed above, $L(x_0) \leq f(x_0) \leq U(x_0)$, hence $U(x_0) = f(x_0) = L(x_0)$. It remains to be shown that f has a limit at x_0 and that $\lim_{x \to x_0} f(x) = f(x_0)$. Choose $\epsilon > 0$. Now $L(x_0) - \epsilon$ is not an upper bound for $\{f(y) : y < x_0\}$, and $U(x_0) + \epsilon$

is not a lower bound for $\{f(y) : x_0 < y\}$; hence there are real numbers y_1 and y_2 such that $\alpha \leq y_1 < x_0 < y_2 \leq \beta$ and such that $L(x_0) - \epsilon < f(y_1)$ and $f(y_2) < U(x_0) + \epsilon$. Let $\delta = \min\{x_0 - y_1, y_1 - x_0\}$. Now if $0 < |x - x_0| < \delta$, then $y_1 < x < y_2$; hence

$$f(x_0) - \epsilon = L(x_0) - \epsilon < f(y_1) \leq f(x) \leq f(y_2) < U(x_0) + \epsilon$$
$$= f(x_0) + \epsilon;$$

that is, $|f(x) - f(x_0)| < \epsilon$. Thus, f has a limit at x_0 and $\lim_{x\to x_0} f(x) = f(x_0)$.

You have probably observed that we have neglected considering the behavior at α and β. Of course, we cannot define U at β, since if $\beta < y$, y is not in the domain of f; and similarly we cannot define L at α. The proof that f has a limit at α and at β is left as Exercise 24. Equipped with this lemma, we are prepared to prove Theorem 2.8.

2.8 THEOREM Let $f: [\alpha, \beta] \to R$ be monotone. Then $D = \{x : x \in (\alpha, \beta)$ and f does not have a limit at $x\}$ is countable. If f has a limit at $x_0 \in (\alpha, \beta)$, then $\lim_{x\to x_0} f(x) = f(x_0)$.

Proof. Assume f is increasing.
Let $D = \{x : x \in (\alpha, \beta)$ and f does not have a limit at $x_0\}$.
Now by Lemma 2.7, $x \in D$ iff $U(x) - L(x) \neq 0$ and, since f is increasing, iff $U(x) - L(x) > 0$. Let

$$D_n = \left\{x : U(x) - L(x) > \frac{1}{n}\right\}.$$

It is clear that $D = \bigcup_{n=1}^{\infty} D_n$. The proof will be complete if we can show that each D_n is finite. Suppose $\{x_1, \ldots, x_r\} \subset D_n$ with $\alpha < x_1 < x_2 < \cdots < x_r < \beta$. Choose $z_1, \ldots, z_{r+1}$ such that $\alpha < z_1 < x_1$, $x_i < z_{i+1} < x_{i+1}$ for $i = 1, 2, \ldots, r - 1$, and $x_r < z_{r+1} < \beta$. Now for each i, $f(z_i) \leq L(x_i)$ and $U(x_i) \leq f(z_{i+1})$; hence

$$f(z_{i+1}) - f(z_i) \geq U(x_i) - L(x_i) > \frac{1}{n}.$$

Now

$$f(\beta) - f(\alpha) = f(\beta) - f(z_{r+1})$$
$$\sum_{k=2}^{r+1} [f(z_k) - f(z_{k-1})] + f(z_1) - f(\alpha) \geq r\left(\frac{1}{n}\right).$$

Since $f(\beta) - f(\alpha) > 0$ is a fixed real number, it is necessary that $r \leq n[f(\beta) - f(\alpha)]$. Therefore, D_n is finite for each n; hence, $D = \bigcup_{n=1}^{\infty} D_n$ is countable.

Note that if f is decreasing, then $-f$ is increasing, and f has a limit at x_0 iff $-f$ has a limit at x_0. In particular, a lemma similar to Lemma 2.7 may be deduced for decreasing functions. See Exercise 23.

Reflect a bit on the content of this theorem. With only the assumption that $f: [\alpha, \beta] \to R$ is monotone, it is possible to prove that f has a limit everywhere except at points of a countable set; and except at α and β, wherever the limit exists, $\lim_{x \to x_0} f(x) = f(x_0)$.

EXERCISES

2.1 DEFINITION OF THE LIMIT OF A FUNCTION

1. Define $f: (-2, 0) \to R$ by $f(x) = \dfrac{x^2 - 4}{x + 2}$. Prove that f has a limit at -2 and find it.
2. Define $f: (-2, 0) \to R$ by $f(x) = \dfrac{2x^2 + 3x - 2}{x + 2}$. Prove that f has a limit at -2 and find it.
3. Give an example of a function $f: (0, 1) \to R$ that has a limit at every point of $(0, 1)$ except $\frac{1}{2}$.
4. Give an example of a function $f: R \to R$ that is bounded and has a limit at every point except -2.

*5. Suppose $f: D \to R$ with x_0 an accumulation point of D. Assume L_1 and L_2 are limits of f at x_0. Prove $L_1 = L_2$. (Use only the definition, since in later theorems, this uniqueness is assumed.)

6. Define $f: (0, 1) \to R$ by $f(x) = \cos \frac{1}{x}$. Does f have a limit at 0? Explain.
7. Define $f: (0, 1) \to R$ by $f(x) = x \cos \frac{1}{x}$. Does f have a limit at 0? Explain.
8. Define $f: (0, 1) \to R$ by $f(x) = \dfrac{x^3 - x^2 + x - 1}{x - 1}$. Prove that f has a limit at 1.
9. Define $f: (-1, 1) \to R$ by $f(x) = \dfrac{x + 1}{x^2 - 1}$. Does f have a limit at 1? Explain.

2.2 LIMITS OF FUNCTIONS AND SEQUENCES

10. Consider $f: (0, 2) \to R$ defined by $f(x) = x^x$. Assume that f has a limit at 0 and find that limit. (*Hint:* Choose a sequence $\{x_n\}_{n=1}^{\infty}$ converging to 0 such that the limit of the sequence $\{f(x_n)\}_{n=1}^{\infty}$ is easy to determine.)

*11. Suppose f, g, and h: $D \to R$ where x_0 is an accumulation point of D, $f(x) \leq g(x) \leq h(x)$ for all $x \in D$, and f and h have limits at x_0 with $\lim_{x \to x_0} f(x) = \lim_{x \to x_0} h(x)$. Prove that g has a limit at x_0 and

$$\lim_{x \to x_0} f(x) = \lim_{x \to x_0} g(x) = \lim_{x \to x_0} h(x).$$

*12. Suppose $f: D \to R$ has a limit at x_0. Prove that $|f| : D \to R$ has a limit at x_0 and that $\lim_{x \to x_0} |f(x)| = |\lim_{x \to x_0} f(x)|$.

13. Define $f: R \to R$ by $f(x) = x - [x]$. (See Example 2.6 for the definition of $[x]$.) Determine those points at which f has a limit and justify your conclusions.

14. Define $f: R \to R$ as follows:

$$f(x) = 8x \text{ if } x \text{ is a rational number}$$
$$f(x) = 2x^2 + 8 \text{ if } x \text{ is an irrational number}$$

Use sequences to guess at which points f has a limit and then use ϵ's and δ's to justify your conclusions.

15. Let $f: D \to R$ with x_0 as an accumulation point of D. Prove that f has a limit at x_0 if for each $\epsilon > 0$, there is a neighborhood Q of x_0 such that, for any $x, y \in Q \cap D, x \neq x_0, y \neq x_0$, we have $|f(x) - f(y)| < \epsilon$.

2.3 ALGEBRA OF LIMITS

16. Define $f: (0, 1) \to R$ by $f(x) = \dfrac{x^3 + 6x^2 + x}{x^2 - 6x}$. Prove that f has a limit at 0 and find that limit.

17. Define $f: R \to R$ as follows:

$$f(x) = x - [x] \text{ if } [x] \text{ is even}$$
$$f(x) = x - [x + 1] \text{ if } [x] \text{ is odd}$$

Determine those points where f has a limit and justify your conclusions.

18. Define $g: (0, 1) \to R$ by $g(x) = \dfrac{\sqrt{1 + x} - 1}{x}$. Prove that g has a limit at 0 and find it.

19. Define $f: (0, 1) \to R$ by $f(x) = \dfrac{\sqrt{9 - x} - 3}{x}$. Prove that f has a limit at 0 and find it.

***20.** Prove Theorem 2.5.

21. Suppose $g: D \to R$ with x_0 an accumulation point of D and $g(x) \neq 0$ for all $x \in D$. Further assume that g has a limit at x_0 and $\lim_{x \to x_0} g(x) \neq 0$. State and prove a theorem similar to Lemma 1.10 for such a function.

22. Show by an example that, even though f and g fail to have limits at x_0, it is possible for $f + g$ to have a limit at x_0. Give similar examples for fg and $\frac{f}{g}$.

2.4 LIMITS OF MONOTONE FUNCTIONS

23. State and prove a lemma similar to Lemma 2.7 for decreasing functions.

24. Let $f: [a, b] \to R$ be monotone. Prove that f has a limit both at a and at b.

25. Suppose $f: [a, b] \to R$ and define $g: [a, b] \to R$ as follows:

$$g(x) = \sup \{f(t): a \leq t \leq x\}.$$

Prove that g has a limit at x_0 if f has a limit at x_0 and $\lim_{t \to x_0} f(t) = f(x_0)$.

MISCELLANEOUS

26. Let f be a function defined on $\{x : x > a\}$. We say that f has a limit at ∞ if there is a number A such that for each $\epsilon > 0$ there is a real number M such that, if $y > M$ and $y > a$, then $|f(y) - A| \leq \epsilon$. Prove that the following limits exist and are as indicated.

a. $\displaystyle\lim_{x \to \infty} \frac{\sqrt{x}}{1 - \sqrt{x}} = -1$ b. $\displaystyle\lim_{x \to \infty} \frac{\sqrt{x}}{x + 2} = 0.$

27. Suppose $f: [a, b] \to R$ and $x_0 \in [a, b]$.
   a. Give a precise and reasonable definition of the statement "the limit of f at x_0 is ∞."
   b. Give an example of a function that satisfies your definition in (a).

28. Assume that $f: R \to R$ is such that $f(x + y) = f(x)f(y)$ for all $x, y \in R$. If f has a limit at zero, prove that f has a limit at every point and either $\lim_{x \to x_0} f(x) = 1$ or $f(x) = 0$ for all $x \in R$.

29. Suppose $f: D \to R$, $g: E \to R$, x_0 is an accumulation point of $D \cap E$ and there is $\epsilon > 0$ such that $D \cap [x_0 - \epsilon, x_0 + \epsilon] = E \cap [x_0 - \epsilon, x_0 + \epsilon]$. If $f(x) = g(x)$ for all $x \in D \cap E \cap [x_0 - \epsilon, x_0 + \epsilon]$, prove that f has a limit at x_0 iff g has a limit at x_0.

PROJECT

The purpose of this project is to ascertain under what conditions an additive function has a limit at each point in R. We say that $f: R \to R$ is *additive* if for all $x, y \in R$, $f(x + y) = f(x) + f(y)$. In what follows, f is an additive function.

1. Show that for each positive integer n and each real number x, $f(nx) = nf(x)$.
2. Suppose f is such that there are $M > 0$ and $a > 0$ such that if $x \in [-a, a]$, then $|f(x)| \leq M$. Choose $\epsilon > 0$. There is a positive integer N such that $\frac{M}{N} < \epsilon$. Show that if $|x - y| < \frac{a}{N}$, then $|f(x) - f(y)| < \epsilon$.
3. Prove that if there are $M > 0$ and $a > 0$ such that if $x \in [-a, a]$, then $|f(x)| \leq M$, then f has a limit at each x in R and $\lim_{t \to x} f(t) = f(x)$.
4. Prove that if f has a limit at each $x \in R$, then there are $M > 0$ and $a > 0$ such that if $x \in [-a, a]$, then $|f(x)| \leq M$.

You have now proven that if $f: R \to R$ is additive, then f has a limit at each point in R iff there are $M > 0$ and $a > 0$ such that if $x \in [-a, a]$, then $|f(x)| \leq M$. In addition, if the condition is satisfied $\lim_{t \to x} f(t) = f(x)$.

See exercise 46 in Chapter 3 for the final word on such functions. For a more elaborate discussion of this problem, see Ralph P. Boas' "A Primer of Real Functions," 3rd Edition, Carus Mathematical Monograph, No. 13, 1981 pp. 125 ff. See Hahn and Rosenthal's *Set Functions*, University of New Mexico Press, 1948, pp. 100 ff. for an additive function that fails to have a limit at any point.

CHAPTER 3

Continuity

In the discussion of the limit of a function it was emphasized that the value of f at x_0 has no bearing on the question of the existence of the limit of a function f at a point x_0. The concept of continuity of a function at a point brings the value of f at x_0 back into the picture. In very imprecise language, a function f will be continuous at a point x_0 if $f(x)$ lies close to $f(x_0)$ whenever x is sufficiently close to x_0. Let us examine this notion in a very familiar setting. Consider the function $f(x) = \sqrt{x}$ for $x \geqslant 0$. Since $f(2)$ is an irrational number, it cannot be expressed by any of the conventional means as a decimal. Nevertheless, for purposes of computation, we can attempt to obtain rational numbers fairly close to $\sqrt{2}$. For example, $(1.4)^2 = 1.96$, $(1.41)^2 = 1.9881$, $(1.414)^2 = 1.999396$, etc. In essence, we are observing that $f(1.96) = 1.4$, $f(1.9881) = 1.41$, and $f(1.999396) = 1.414$, and we assume that 1.414 is close to $\sqrt{2}$ since 1.414 is the value of f at a point close to 2. Thus, the assumption is that $f(x) = \sqrt{x}$ is continuous at 2, a fact to be proved later. As with all new ideas, a precise definition is needed.

3.1 CONTINUITY OF A FUNCTION AT A POINT

DEFINITION Suppose $E \subset R$ and $f\colon E \to R$. If $x_0 \in E$, then f is *continuous* at x_0 iff for each $\epsilon > 0$, there is a $\delta > 0$ such that if

$$|x - x_0| < \delta, \quad x \in E,$$

then

$$|f(x) - f(x_0)| < \epsilon.$$

If f is continuous at x for every $x \in E$, then we say f is continuous.

Compare this definition with the definition in the previous chapter concerning the limit of a function at a point x_0. First of all, for continuity at x_0, the number x_0 must belong to E but it need not be an accumulation point of E.

Indeed, if $f: E \to R$ with $x_0 \in E$ and x_0 not an accumulation point of E, then there is $\delta > 0$ such that if $|x - x_0| < \delta$ and $x \in E$, then $x = x_0$; hence

$$|f(x) - f(x_0)| = 0 < \epsilon$$

for every $\epsilon > 0$. In other words, if x_0 is not an accumulation point of E and $x_0 \in E$, then f is continuous at x_0 by default. Thus, the only interesting case is when x_0 is an accumulation point of E. With this assumption, the following theorem seems the natural conclusion after comparing the two aforementioned definitions and related theorems from Chapter 2.

3.1 THEOREM Let $f: E \to R$ with $x_0 \in E$ and x_0 an accumulation point of E. Then (i)–(iii) below are equivalent:

i. f is continuous at x_0.
ii. f has a limit at x_0 and $\lim_{x \to x_0} f(x) = f(x_0)$.
iii. For every sequence $\{x_n\}_{n=1}^{\infty}$ converging to x_0 with $x_n \in E$ for each n, $\{f(x_n)\}_{n=1}^{\infty}$ converges to $f(x_0)$.

This theorem is the first of its type—that is, one that asserts that two or more statements are equivalent. Our method of proof will be to show that (iii) implies (ii), then show that (ii) implies (i), and finally show that (i) implies (iii). This completes the cycle showing that (i), (ii), and (iii) are equivalent—that is, that each implies and is implied by the other. Also, if x_0 is not an accumulation point of E with $x_0 \in E$, then (i) is always true, as is (iii). Exercise 4 asks you to prove that (iii) is true in this case.

Proof. Assume (iii) holds. In particular, if $\{x_n\}_{n=1}^{\infty}$ converges to x_0 with $x_n \neq x_0$ and $x_n \in E$ for all n, then $\{f(x_n)\}_{n=1}^{\infty}$ converges to $f(x_0)$. Hence, by Theorem 2.1, f has a limit at x_0 and $\lim_{x \to x_0} f(x) = f(x_0)$. Thus, (iii) implies (ii).

Assume (ii) holds and choose $\epsilon > 0$. Since (ii) holds, there is $\delta > 0$ such that if $0 < |x - x_0| < \delta$ and $x \in E$, then

$$|f(x) - f(x_0)| < \epsilon$$

The only fact needed to fulfill the definition of continuity at x_0 is to remove the condition $0 < |x - x_0|$. However, when $|x - x_0| = 0$, then $x = x_0$; hence

$$|f(x) - f(x_0)| = 0 < \epsilon.$$

Thus, f is continuous at x_0; in other words, (ii) implies (i).

Suppose now that (i) holds and that $\{x_n\}_{n=1}^{\infty}$ is a sequence of points in E that converges to x_0. Choose $\epsilon > 0$. There is $\delta > 0$ such that for $|x - x_0| < \delta$ and $x \in E$, $|f(x) - f(x_0)| < \epsilon$. Since $\{x_n\}_{n=1}^{\infty}$ converges to x_0, there is N such that for $n \geqslant N$, $|x_n - x_0| < \delta$. Thus, for $n \geqslant N$, $|f(x_n) - f(x_0)| < \epsilon$. This shows that $\{f(x_n)\}_{n=1}^{\infty}$ converges to $f(x_0)$. (Does this paragraph look familiar? It should. Compare it with the proof of Theorem 2.1.)

It is worthwhile to reflect on this theorem and consider some of its less obvious uses. First of all, it may be used to show that a given function f is not continuous at a point x_0 by exhibiting a sequence $\{x_n\}_{n=1}^{\infty}$ converging to x_0 where the sequence $\{f(x_n)\}_{n=1}^{\infty}$ does not converge at $f(x_0)$. From a slightly different point of view, this theorem may be used to show that a certain sequence is convergent. For example, assume that the function $f(x) = e^x$ is continuous on R. Thus, in particular, if the sequence $\{x_n\}_{n=1}^{\infty}$ converges to x_0, then the continuity of $f(x) = e^x$ and Theorem 3.1 guarantee that $\{e^{x_n}\}_{n=1}^{\infty}$ converges to e^{x_0}. We shall call attention to this aspect of Theorem 3.1 later.

We pause momentarily to reconsider some of the results presented in Chapter 2. Recall the function $f\colon R \to R$ defined by

$$f(x) = \frac{x^2 - 1}{x - 1}$$

for $x \neq 1$ and $f(1) = 6$. It has been shown that f has a limit at 1 and $\lim_{x \to 1} f(x) = 2$. Unfortunately, $\lim_{x \to 1} f(x) = 2 \neq 6 = f(1)$; hence, f is not continuous at 1. In this case, f fails to be continuous at 1 because $f(1)$ does not happen to be the limit of f at 1. Now consider the function $g(x) = \sin \frac{1}{x}$ for $0 < x < 1$, and $g(0) = 38$. We have seen that g fails to have a limit at zero, so g is not continuous at zero, regardless of what we try for $g(0)$. The function $h(x) = \frac{1}{x}$ for $0 < x < 1$ behaves similarly in the sense that h doesn't have a limit at zero, so it is futile to attempt to define $h(0)$ such that h will be continuous at zero.

Recall the function $f\colon [0, 1] \to R$ defined such that $f(x) = 0$ if x is irrational and $f\left(\frac{p}{q}\right) = \frac{1}{q}$ if p and q are positive integers that are relatively prime. It was shown in Chapter 2 that f has a limit at each point of $[0, 1]$ and that its limit is zero at each point. Thus, f is continuous at $x \in [0, 1]$ iff $f(x) = 0$ (that is, iff x is irrational).

3.2 ALGEBRA OF CONTINUOUS FUNCTIONS

The next theorem is a natural follow-up on Theorem 2.4.

3.2 THEOREM Suppose $f\colon D \to R$ and $g\colon D \to R$ are continuous at $x_0 \in D$. Then

1. $f + g$ is continuous at x_0.
2. fg is continuous at x_0.
3. If $g(x_0) \neq 0$, $\frac{f}{g}$ is continuous at x_0.

[A word of advice is necessary concerning (3). The function $\frac{f}{g}$ is defined only where $g(x) \neq 0$; hence, when speaking of the function $\frac{f}{g}$, we assume that the domain consists of those points where $g(x) \neq 0$. Lemma 3.3 will shed some light on this situation.]

Proof. We shall use a different method of proof for each part of this theorem in order to point out the use of facts previuosly presented on sequences and limits and also to show a direct proof.

Suppose f and g are continuous at $x_0 \in D$. Let $\{x_n\}$ be any sequence of points in D that converges to x_0. Then, by Theorem 3.1, $\{f(x_n)\}$ converges to $f(x_0)$ and $\{g(x_n)\}_{n=1}^{\infty}$ converges to $g(x_0)$; hence

$$\{f(x_n) + g(x_n)\}_{n=1}^{\infty}$$

converges to $f(x_0) + g(x_0)$. Thus,

$$\{(f + g)(x_n)\}_{n=1}^{\infty} = \{f(x_n) + g(x_n)\}_{n=1}^{\infty}$$

converges to $f(x_0) + g(x_0) = (f + g)(x_0)$, and hence, by Theorem 3.1, $f + g$ is continuous at x_0. This concludes the proof of (1).

Suppose f and g are continuous at $x_0 \in D$. If x_0 is not an accumulation point of D, then fg is continuous at x_0 by our observations following the definition of continuity. Suppose now that x_0 is an accumulation point of D. By Theorem 3.1, f and g have limits at x_0 and

$$\lim_{x \to x_0} f(x) = f(x_0) \quad \text{and} \quad \lim_{x \to x_0} g(x) = g(x_0).$$

By Theorem 2.4, fg has a limit at x_0 and

$$\begin{aligned} \lim_{x \to x_0} (fg)(x) &= [\lim_{x \to x_0} f(x)][\lim_{x \to x_0} g(x)] = f(x_0)g(x_0) \\ &= (fg)(x_0). \end{aligned}$$

Thus, by Theorem 3.1, fg is continuous at x_0.

We have purposely elected to give a direct proof of (3) because it involves some important facts not yet brought to light. We hope the reader has attempted a direct proof of Theorem 2.4 (3) and discovered the need for a lemma similar to the following. The proof of Theorem 3.2 (3) follows this lemma.

3.3 LEMMA Let $g\colon D \to R$ be continuous at $x_0 \in D$ with $g(x_0) \neq 0$. Then there are $\delta > 0$ and $\alpha > 0$ such that if $|x - x_0| < \delta$ and $x \in D$, then $|g(x)| \geqslant \alpha$.

(Note the similarity between this lemma and Lemma 1.10, the one needed for the theorem concerning the convergence of the quotient of two convergent sequences.)

Proof. Choose

$$\alpha = \frac{|g(x_0)|}{2} > 0.$$

There is $\delta > 0$ such that for $|x - x_0| < \delta$, $x \in D$, we have

$$|g(x) - g(x_0)| < \alpha.$$

Thus,

$$|g(x)| \geq |g(x_0)| - |g(x) - g(x_0)| > |g(x_0)| - \alpha = \frac{|g(x_0)|}{2} = \alpha.$$

Proof of 3.2 continued. Suppose g is continuous at x_0 and $g(x_0) \neq 0$. By Lemma 3.3, there are $\delta > 0$ and $\alpha > 0$ such that for $|x - x_0| < \delta$ and $x \in D$, we have $|g(x)| \geq \alpha$. Choose $\epsilon > 0$. Let

$$\epsilon' = \alpha|g(x_0)|\epsilon > 0.$$

There is $\delta' > 0$ such that for $|x - x_0| < \delta'$ and $x \in D$,

$$|g(x) - g(x_0)| < \epsilon'.$$

Let

$$\delta'' = \min\{\delta', \delta\}.$$

Now $\delta'' > 0$ and for $|x - x_0| < \delta''$ with $x \in D$, we have

$$\left|\frac{1}{g}(x) - \frac{1}{g}(x_0)\right| = \left|\frac{g(x_0) - g(x)}{g(x)g(x_0)}\right| < \frac{\epsilon'}{\alpha|g(x_0)|} = \epsilon.$$

Thus, $\frac{1}{g}$ is continuous at x_0. By part (2) of this theorem, $\frac{f}{g}$ is then continuous at x_0.

From our remarks in Chapter 2 and Theorem 3.1, it is clear that every polynomial is continuous. Furthermore, if p and q are polynomials, then $\frac{p}{q}$ is continuous at any point x where $q(x) \neq 0$. The case where $q(x) = 0$ is handled in detail in Exercise 11 at the end of this chapter.

■ **Example 3.1** The function $f(x) = \sin x$ is undoubtedly familiar to the reader. We observe that, for $0 \leq x \leq \frac{\pi}{2}$, $0 \leq \sin x \leq x$. From this and the fact that $\sin(-x) = -\sin x$, it is clear that $f(x) = \sin x$ has a limit at zero, namely, $\lim_{x \to 0} \sin x = 0$. From this, we can deduce that the sine function is continuous at every point. Now for all x and y,

$$\sin x - \sin y = 2\cos\left(\frac{x + y}{2}\right)\sin\left(\frac{x - y}{2}\right).$$

Let x_0 be any real number and $\epsilon > 0$. Let $\delta = \min\{\epsilon, \pi\}$. Then for $|x_0 - y| < \delta$,

$$|\sin x_0 - \sin y| = \left|2 \cos\left(\frac{x_0 + y}{2}\right) \sin\left(\frac{x_0 - y}{2}\right)\right| \leqslant 2\left|\frac{x_0 - y}{2}\right|$$
$$= |x_0 - y| < \delta \leqslant \epsilon.$$

Thus, $\sin x$ is continuous at x_0. For all real x, $\cos x = \sin\left(\frac{\pi}{2} - x\right)$. In order to verify the continuity of $\cos x$ at x_0, let us proceed directly. Choose $\epsilon > 0$ and let $\delta = \min\{\epsilon, \pi\}$. If $|x - x_0| < \delta$, then

$$\left|\left(\frac{\pi}{2} - x\right) - \left(\frac{\pi}{2} - x_0\right)\right| = |x_0 - x| < \delta;$$

hence

$$|\cos x - \cos x_0| = \left|\sin\left(\frac{\pi}{2} - x\right) - \sin\left(\frac{\pi}{2} - x_0\right)\right| < \delta \leqslant \epsilon.$$

Thus, $\cos x$ is continuous at x_0. Now it is clear that the functions $\tan x$, $\sec x$, $\csc x$, and $\operatorname{ctn} x$ are continuous where defined, since they may be defined by appropriate products and quotients of the functions $\sin x$ and $\cos x$. ■

Two points that come to light in this last paragraph deserve some special attention. First of all, a function $f\colon D \to R$ is continuous at a point $x_0 \in D$ iff for each $\epsilon > 0$, there is $\delta > 0$ such that for $|x - x_0| < \delta$ and $x \in D$, we have $|f(x) - f(x_0)| < \epsilon$. Note that the choice of $\delta > 0$ is influenced by two things: the choice of $\epsilon > 0$ and the point in question, x_0. In the process of investigating the continuity of $f(x) = \sin x$, it was discovered that if $\epsilon > 0$ is given, $\delta > 0$ could be found that would be suitable for any x_0. This is not the usual state of affairs and will be considered in some detail very soon. Secondly, the continuity of the cosine function was deduced from the facts that the sine function is continuous and that $\cos x = \sin\left(\frac{\pi}{2} - x\right)$. If we let $g(x) = \cos x$, $f(x) = \sin x$, and $h(x) = \frac{\pi}{2} - x$, then this identity may be written $g = f \circ h$. Of course, since h is a polynomial function, it is continuous; indeed, that fact was used to observe that if $|x - x_0| < \delta$, then

$$|h(x) - h(x_0)| = \left|\left(\frac{\pi}{2} - x\right) - \left(\frac{\pi}{2} - x_0\right)\right| = |x - x_0| < \delta.$$

We seek now to generalize this idea and shall attempt to prove that the composition of continuous functions is continuous.

3.4 THEOREM If $f\colon D \to R$ and $g\colon D' \to R$ with $\operatorname{im} f \subset D'$, where f is continuous at $x_0 \in D$ and g is continuous at $f(x_0)$, then $g \circ f$ is continuous at x_0.

Proof. Choose $\epsilon > 0$. There is $\delta_1 > 0$ such that if $|y - f(x_0)| < \delta_1$ and $y \in D'$, then

$$|g(y) - g(f(x_0))| < \epsilon.$$

There is $\delta_2 > 0$ such that if $|x - x_0| < \delta_2$ and $x \in D$, then

$$|f(x) - f(x_0)| < \delta_1.$$

Now, since im $f \subset D'$, if $|x - x_0| < \delta_2$ and $x \in D$, then $f(x) \in D'$, $f(x_0) \in D'$, and $|f(x) - f(x_0)| < \delta_1$; hence,

$$|(g \circ f)(x) - (g \circ f)(x_0)| = |g(f(x)) - g(f(x_0))| < \epsilon.$$

Therefore, $g \circ f$ is continuous at x_0.

This theorem, coupled with previous results, allows us to conclude immediately that functions such as $f(x) = \cos x^2$ and $g(x) = \sin(\cos x)$ are continuous at x_0 for all real numbers x_0.

Let us now take up the remaining topics that we want to pursue further. We shall give the appropriate definition and then consider some examples.

3.3 UNIFORM CONTINUITY: OPEN, CLOSED, AND COMPACT SETS

DEFINITION A function $f: D \to R$ is *uniformly continuous on* $E \subset D$ iff for every $\epsilon > 0$, there is $\delta > 0$ such that if $x, y \in E$ with $|x - y| < \delta$, then $|f(x) - f(y)| < \epsilon$. If f is uniformly continuous on D, we say f is uniformly continuous.

As promised, some examples and nonexamples are forthcoming.

■ **Example 3.2** Consider the function f: $[2.5, 3] \to R$ defined by $f(x) = \dfrac{3}{x - 2}$. We will show that f is uniformly continuous on $[2.5, 3]$. Choose $\epsilon > 0$. Now

$$f(x) - f(y) = \frac{3}{x - 2} - \frac{3}{y - 2} = \frac{3(y - x)}{(x - 2)(y - 2)}.$$

We see no difficulty making the numerator of this last fraction small, but, as we have seen before, the problem of bounding the denominator away from zero must be addressed. However, since x and y must belong to $[2.5, 3]$, the smallest nonnegative value for the denominator is 0.25, occurring when $x = y = 2.5$. With this in mind, we choose $\delta = \frac{1}{12}\epsilon$. Thus, if $|x - y| < \delta$ and $x, y \in [2.5, 3]$, then

$$|f(x) - f(y)| = \left|\frac{3(y - x)}{(x - 2)(y - 2)}\right| \leqslant \frac{|3(x - y)|}{0.25} < 12\delta = \epsilon. \quad ■$$

■ **Example 3.3** Consider the function $f\colon (0, 6) \to R$ defined by $f(x) = x^2 + 2x - 5$. We will show that f is uniformly continuous on $(0, 6)$. Choose $\epsilon > 0$. Now

$$\begin{aligned}|f(x) - f(y)| &= |(x^2 + 2x - 5) - (y^2 + 2y - 5)| = |(x^2 - y^2) - 2(x - y)| \\ &= |(x - y)(x + y + 2)|\end{aligned}$$

To make this expression small, we need to make $|x - y|$ small and *bound* $|x + y + 2|$. Although 6 is not in the domain of f, we can use that number to bound $|x + y + 2|$, that is, $|x + y + 2| < 6 + 6 + 2 = 14$. With this in mind, we choose $\delta = \frac{1}{14}\epsilon$. Thus, if $|x - y| < \delta$ and $x, y \in (0, 6)$, then

$$|f(x) - f(y)| = |(x - y)(x + y + 2)| < 14\delta = \epsilon. \quad ■$$

The sine function has been shown to be uniformly continuous on the set of all real numbers. Therefore, it is a uniformly continuous function. Similarly, the cosine function is uniformly continuous. This is not surprising in light of Exercise 23.

Now consider the function $f\colon (0, \frac{\pi}{2}) \to R$ defined by $f(x) = \tan x$. Since $\tan x = \dfrac{\sin x}{\cos x}$, for $0 < x < \frac{\pi}{2}$, f is a quotient of uniformly continuous functions. [Since $\sin x$ and $\cos x$ are uniformly continuous on R, they are certainly uniformly continuous on $(0, \frac{\pi}{2})$.] Exercises 18 and 19 show under what conditions the sum, product, or composition of uniformly continuous functions yields a uniformly continuous function. We know that the tangent function is unbounded near $\frac{\pi}{2}$, so it cannot have a limit there. Let us show that this guarantees that $\tan x$ is not uniformly continuous on $(0, \frac{\pi}{2})$.

3.5 THEOREM Let $f\colon D \to R$ be uniformly continuous. Then, if x_0 is an accumulation point of D, f has a limit at x_0.

Proof. Let x_0 be an accumulation point of D, and let $\{x_n\}_{n=1}^{\infty}$ be any sequence of members of $D \setminus \{x_0\}$ converging to x_0. Recall that it is sufficient to prove that $\{f(x_n)\}_{n=1}^{\infty}$ is a Cauchy sequence, since every Cauchy sequence of real numbers is convergent. Choose $\epsilon > 0$. Since f is uniformly continuous on D, there is $\delta > 0$ such that for all $x, y \in D$, $|x - y| < \delta$ implies $|f(x) - f(y)| < \epsilon$. Since $\{x_n\}_{n=1}^{\infty}$ converges to x_0, it is a Cauchy sequence, and there is N such that for $m, n \geqslant N$, $|x_n - x_m| < \delta$. Now $x_n \in D$ for each n; hence, for $m, n \geqslant N$, $|x_n - x_m| < \delta$ and $x_m, x_n \in D$; hence

$$|f(x_n) - f(x_m)| < \epsilon.$$

Thus, $\{f(x_n)\}_{n=1}^{\infty}$ is Cauchy. By Theorem 2.1, f has a limit at x_0.

■ **Example 3.4** Observe that Theorem 3.5 gives a necessary condition for uniform continuity, but not one that is sufficient. For example, consider $g\colon R \to R$ where $g(x) = x^2$ for all real numbers x. Since g is continuous and dom $g = R$, the function g has a limit at every accumulation point of R because R contains all its accumulation points. However,

$$|g(x) - g(y)| = |x^2 - y^2| = |x + y|\,|x - y|.$$

Choose $\epsilon > 0$ and consider any $\delta > 0$. Now choose x and y such that $|x - y| = \dfrac{\delta}{2} < \delta$ and $|x + y| = \dfrac{3\epsilon}{\delta}$. Then

$$|g(x) - g(y)| = |x + y|\,|x - y| = \frac{3\epsilon}{\delta}\frac{\delta}{2} = \frac{3}{2}\epsilon > \epsilon.$$

In effect, given $\epsilon > 0$, it is impossible to find $\delta > 0$ such that for $|x - y| < \delta$, we always have $|g(x) - g(y)| < \epsilon$. Thus, g is not uniformly continuous. ■

We have seen two examples of continuous functions that were not uniformly continuous; in the first case, the domain was $(0, \frac{\pi}{2})$—a set with accumulation points that did not belong to the set—and the second function had an unbounded domain. The difficulty in the first example comes about near the accumulation points and perhaps may be overcome by considering functions whose domains contain all their accumulation points. Such sets are of sufficient importance to deserve a name.

DEFINITION A set $E \subset R$ is *closed* iff every accumulation point of E belongs to E.

As has been observed previously, a finite set has no accumulation points and, by definition, is a closed set. Also, R, the set of all real numbers, is a closed set, since an accumulation point of R is necessarily a real number, hence a member of R. Suppose E is a closed set and $x_0 \in R \setminus E$. Then, by the definition of a closed set, x_0 is not an accumulation point of E; hence, there is a neighborhood Q of x_0 that contains no points of E (since $x_0 \not\in E$), so $Q \subset R \setminus E$. Thus, if E is closed, there is, for each $x_0 \in R \setminus E$, a neighborhood Q of x_0 such that $Q \subset R \setminus E$. This special type of set comes about in a natural way by considering the complement of a closed set.

DEFINITION A set $A \subset R$ is *open* iff for each $x \in A$ there is a neighborhood Q of x such that $Q \subset A$.

The remarks leading up to this definition yield the proof of half of the following theorem.

3.6 THEOREM A set $E \subset R$ is closed iff $R \setminus E$ is open.

Proof. Suppose E is closed and $x_0 \in R \setminus E$. Then, since E contains all its accumulation points, x_0 is not an accumulation point of E. Hence, there is a neighborhood Q of x_0 that contains no points of E. (Recall that $x_0 \not\in E$.) But now $Q \subset R \setminus E$, hence (by definition), $R \setminus E$ is open.

Suppose $R \setminus E$ is open. To show E is closed, it sufficies to show that if x_0 is an accumulation point of E, then $x_0 \in E$. Let x_0 be an accumulation point of E. Then $x_0 \in E$ or $x_0 \in R \setminus E$. However, $R \setminus E$ is open; so if $x_0 \in R \setminus E$, there is a neighborhood Q of x_0 such that $Q \subset R \setminus E$ or $Q \cap E$ is empty, contrary to x_0 being an accumulation point of E. Thus, $x_0 \in E$ and E is closed.

Now let us return to the problem that precipitated this discussion. Consider $f\colon E \to R$ with f continuous on E. Choose $\epsilon > 0$. Now for each $x \in E$, there is $\delta_x > 0$ such that if $y \in E$ and $|x - y| < \delta_x$, then $|f(x) - f(y)| < \epsilon$. In the search for $\delta > 0$ to guarantee that $|x - y| < \delta$ and $x, y \in E$ imply that $|f(x) - f(y)| < \epsilon$, one might consider

$$\delta = \inf \{\delta_x : x \in E\}.$$

If E is finite, δ is positive; but if E is infinite, δ might be zero. Indeed, if f is not uniformly continuous, it will be zero. Consider now the set $(x - \delta_x, x + \delta_x) \cap E$. If

$$y_1, y_2 \in (x - \delta_x, x + \delta_x) \cap E,$$

then $|x - y_1| < \delta_x$ and $|x - y_2| < \delta_x$; hence

$$\begin{aligned} |f(y_1) - f(y_2)| &\leq |f(y_1) - f(x) + f(x) - f(y_2)| \\ &\leq |f(y_1) - f(x)| + |f(x) - f(y_2)| < \epsilon + \epsilon = 2\epsilon. \end{aligned}$$

If one could find $x_1, \ldots, x_n$ such that

$$E \subset \bigcup_{i=1}^{n} (x_i - \delta_{x_i}, x_i - \delta_{x_i}),$$

there might be some hope of showing that f is uniformly continuous. Note that for each x, $(x - \delta_x, x + \delta_x)$ is an open set (see Exercise 25), and $E \subset \bigcup_{x \in E} (x - \delta_x, x + \delta_x)$. Here we have a family of open sets whose union contains E, and we wish to choose a finite subfamily with the same property. A new definition is now in order.

DEFINITION A set E is *compact* iff, for every family $\{G_\alpha\}_{\alpha \in A}$ of open sets such that $E \subset \bigcup_{\alpha \in A} G_\alpha$, there is a finite set $\{\alpha_1, \ldots, \alpha_n\} \subset A$ such that $E \subset \bigcup_{i=1}^{n} G_{\alpha i}$.

We shall illustrate this concept by giving some examples of sets that are not compact.

■ **Example 3.5** Let $E = (0, 1]$ and for each positive integer n, let $G_n = (\frac{1}{n}, 2)$. If $0 < x \leq 1$, there is a positive integer n such that $\frac{1}{n} < x$; hence $x \in G_n$, and thus

$$E \subset \bigcup_{n=1}^{\infty} G_n.$$

If we choose a finite set $n_1, \ldots, n_r$ of positive integers, then

$$\bigcup_{i=1}^{r} G_{n_i} = G_{n_0}$$

where $n_0 = \max\{n_1, \ldots, n_r\}$ and

$$E \not\subset G_{n_0} = \left(\frac{1}{n_0}, 2\right).$$

Thus, we have a family of open sets $\{G_n\}_{n \in J}$ such that $E \subset \bigcup_{n \in J} G_n$, but no finite subfamily has this property. From the definition, it is clear that E is not compact. ■

■ **Example 3.6** As a second example, consider the set J of all positive integers. For each $n \in J$, define

$$G_n = \left(n - \frac{1}{2}, n + \frac{1}{2}\right).$$

Now $J \subset \bigcup_{n=1}^{\infty} G_n$, but each G_n contains exactly one member of J, so it is impossible to choose $n_1, \ldots, n_r$ such that $J \subset \bigcup_{i=1}^{r} G_{n_i}$. Thus, J is not compact. ■

Let us introduce some new terminology to help simplify our language. If E is a set and $\{G_\alpha\}_{\alpha \in A}$ is a collection of sets such that $E \subset \bigcup_{\alpha \in A} G_\alpha$, then the collection $\{G_\alpha\}_{\alpha \in A}$ is called a *cover* of E. If each G_α is an open set, then the collection $\{G_\alpha\}_{\alpha \in A}$ is called an *open cover* of E. If $\{G_\alpha\}_{\alpha \in A}$ is a cover of E and $B \subset A$ such that $E \subset \bigcup_{\alpha \in B} G_\alpha$, then the collection $\{G_\alpha\}_{\alpha \in B}$ is called a *subcover* of E; and, if B is finite, then $\{G_\alpha\}_{\alpha \in B}$ is called a *finite subcover* of E. In a sense, we are abusing the language, since every subcover of E is also a cover of E; but in light of the notion of compactness, this is a reasonable abuse. With these additions to our vocabulary, we can restate our definition of compactness: A set E is compact iff every open cover of E has a finite subcover.

The reader should keep the preceding two examples clearly in mind. The first example, $(0, 1]$, is a bounded set that is not closed since 0 is an accumulation point of $(0, 1]$ that does not belong to $(0, 1]$. The second example is a closed set but is unbounded. As the reader might suspect from our choice of examples, the compact sets on the line are precisely those that are both closed and bounded. This is the content of the next theorem.

3.7 HEINE–BOREL THEOREM A set $E \subset R$ is compact iff E is closed and bounded.

Proof. Suppose that E is closed and bounded and that $\{G_\lambda\}_{\lambda \in A}$ is an open cover of E. Since E is bounded, there is a closed interval $[\alpha, \beta]$ such that $E \subset [\alpha, \beta]$. Let us suppose that $\{G_\lambda\}_{\lambda \in A}$ has no finite subcover of E (E is not compact). We shall show that this assumption is self-contradictory. Let

γ_0 be the midpoint of $[\alpha, \beta]$. At least one of the two sets $[\alpha, \gamma_0] \cap E$ and $[\gamma_0, \beta] \cap E$ cannot be covered by a finite subfamily of $\{G_\lambda\}_{\lambda \in A}$. Choose one and call it $[\alpha_1, \beta_1]$; let γ_1 be the midpoint of $[\alpha_1, \beta_1]$. Again at least one of the two sets $[\alpha_1, \gamma_1] \cap E$ and $[\gamma_1, \beta_1] \cap E$ cannot be covered by a finite subfamily of $\{G_\lambda\}_{\lambda \in A}$; choose one such and call it $[\alpha_2, \beta_2]$. (Does this game sound familiar? See the proof of the Bolzano–Weierstrass Theorem.) Continuing in this fashion, we obtain a sequence of closed intervals $[\alpha_n, \beta_n]$ with the following properties:

1. $\beta_n - \alpha_n = \dfrac{1}{2^n}(\beta - \alpha)$.
2. $[\alpha_{n+1}, \beta_{n+1}] \subset [\alpha_n, \beta_n]$ for all n.
3. For each n, the set $[\alpha_n, \beta_n] \cap E$ cannot be covered by a finite subfamily of $\{G_\lambda\}_{\lambda \in A}$.

By (3), $[\alpha_n, \beta_n] \cap E$ is nonempty for each n; hence, we may choose $x_n \in [\alpha_n, \beta_n] \cap E$ for $n = 1, 2, \ldots$. Consider the set

$$P = \{x_n : n = 1, 2, \ldots\}.$$

If P is finite then by (2), there is n_0 such that

$$x_{n0} \in [\alpha_n, \beta_n] \cap E$$

for all n. Since $\{G_\lambda\}_{\lambda \in A}$ is an open cover of E,

$$x_{n0} \in G_{\lambda 0}$$

for some $\lambda_0 \in A$; and since G_{λ_0} is open, there is $\epsilon > 0$ such that

$$(x_{n0} - \epsilon, x_{n0} + \epsilon) \subset G_{\lambda 0}.$$

Choose n such that

$$\beta_n - \alpha_n = \frac{1}{2^n}(\beta - \alpha) < \epsilon.$$

Then, since $x_{n0} \in [\alpha_n, \beta_n] \cap E$,

$$x \in [\alpha_n, \beta_n] \cap E$$

implies that $|x - x_{n0}| \leq \beta_n - \alpha_n < \epsilon$. Hence

$$[\alpha_n, \beta_n] \cap E \subset (x_{n0} - \epsilon, x_{n0} + \epsilon) \subset G_{\lambda 0}.$$

This means that $[\alpha_n, \beta_n] \cap E$ may be covered by a finite subfamily of $\{G_\lambda\}_{\lambda \in A}$, which is contrary to (3).

Suppose P is infinite. Then P is an infinite bounded set and has an accumulation point x_0; and since $P \subset E$, clearly x_0 is an accumulation point of E and $x_0 \in E$ since E is assumed closed. There is $\lambda_1 \in A$ such that $x_0 \in G_{\lambda_1}$; and, since G_{λ_1} is open, there is $\epsilon > 0$ such that

$$(x_0 - \epsilon, x_0 + \epsilon) \subset G_{\lambda_1}.$$

Choose n such that

$$\beta_n - \alpha_n = \frac{1}{2^n}(\beta - \alpha) < \frac{\epsilon}{2}.$$

Now x_0 is an accumulation point of P and $\left(x_0 - \frac{\epsilon}{2}, x_0 + \frac{\epsilon}{2}\right)$ is a neighborhood of x_0. Hence there are infinitely many members of P that belong to $\left(x_0 - \frac{\epsilon}{2}, x_0 + \frac{\epsilon}{2}\right)$. In particular, there is $m > n$ such that

$$x_m \in \left(x_0 - \frac{\epsilon}{2}, x_0 + \frac{\epsilon}{2}\right).$$

Now if $z \in [\alpha_m, \beta_m] \cap E$, then

$$\begin{aligned} |z - x_0| \leqslant |z - x_m| + |x_m - x_0| &< (\beta_m - \alpha_m) + \frac{\epsilon}{2} \\ &\leqslant (\beta_n - \alpha_n) + \frac{\epsilon}{2} < \frac{\epsilon}{2} + \frac{\epsilon}{2} = \epsilon. \end{aligned}$$

Thus, $[\alpha_m, \beta_m] \cap E \subset (x_0 - \epsilon, x_0 + \epsilon) \subset G_{\lambda_1}$, contrary to (3).

We have now arrived at a contradiction to our original assumption—namely, that E was closed and bounded but not compact. Thus, if E is closed and bounded, E is compact.

It now remains to show that every compact set is closed and bounded. We shall accomplish this by showing that if a set is unbounded, it cannot be compact, and if a set is not closed, it cannot be compact.

Assume E is not bounded. For each positive integer n, let $G_n = (-n, n)$. Now G_n is open for each n and $E \subset \bigcup_{n=1}^{\infty} G_n$, so $\{G_n\}_{n=1}^{\infty}$ is an open cover of E. If $n_1, \ldots, n_r$ is any finite set of positive integers, then

$$\bigcup_{i=1}^{r} G_{n_i} = G_{n_0}$$

where $n_0 = \max\{n_1, \ldots, n_r\}$, and, since E is unbounded,

$$\{G_{n_i}\}_{i=1,\ldots,r}$$

is not a cover of E. Thus, $\{G_n\}_{n=1}^{\infty}$ is an open cover of E with no finite subcover; hence E is not compact.

Assume E is not closed. Then there is an accumulation point of E (call it x_0) such that $x_0 \neq E$. In much the same way that we did for the set $(0, 1]$, we shall construct an open cover of E with no finite subcover. For each positive integer n, define

$$G_n = R \setminus \left[x_0 - \frac{1}{n}, x_0 + \frac{1}{n}\right].$$

Now $\left[x_0 - \frac{1}{n}, x_0 + \frac{1}{n}\right]$ is a closed set for each n, hence G_n is open. Observe that $\bigcup_{n=1}^{\infty} G_n = R \setminus \{x_0\}$, so $\{G_n\}_{n=1}^{\infty}$ is an open cover of E. Let $n_1, \ldots, n_r$ be any finite set of positive integers and $n_0 = \max\{n_1, \ldots, n_r\}$. Then

$$\bigcup_{i=1}^{r} G_{n_i} = G_{n_0} = R \setminus \left[x_0 - \frac{1}{n_0}, x_0 + \frac{1}{n_0}\right].$$

In order that $\{G_{ni}\}_{i=1,\ldots,r}$ be a cover of E, we must have

$$E \subset G_{n_0} = R \setminus \left[x_0 - \frac{1}{n_0}, x_0 + \frac{1}{n_0}\right];$$

hence, $E \cap \left(x_0 - \frac{1}{n_0}, x_0 + \frac{1}{n_0}\right)$ is empty, contrary to x_0 being an accumulation point of E. Thus, $\{G_n\}_{n=1}^{\infty}$ has no finite subcover and E is not compact.

Having read the proof of this theorem, you should go back and see what makes it work and where the hypotheses are used. This time, don't worry about the details. Just seek out the rough idea of the proof.

In passing, we give some examples of compact sets. Every finite set is compact since it is necessarily closed and bounded. If $a < b$, then $[a, b]$ is closed and bounded and thus is compact. It is easy to show that the union of finitely many compact sets is compact, so any union of a finite number of closed, bounded intervals is compact. See Exercise 34.

With this knowledge of compact sets in mind, we shall exploit this concept with respect to continuity. The remarks preceding the definition of compactness should make the next theorem no surprise.

3.8 THEOREM Let $f: D \to R$ be continuous with D compact—that is, closed and bounded. Then f is uniformly continuous.

Proof. Choose $\epsilon > 0$. Since f is continuous on D, f is continuous at x for each $x \in D$. Thus, for each $x \in D$, there is $\delta_x > 0$ such that if $|x - y| < \delta_x$ and $y \in D$, then $|f(x) - f(y)| < \frac{\epsilon}{2}$. Consider this family

$$\left\{\left(x - \frac{\delta_x}{2}, x + \frac{\delta_x}{2}\right)\right\}_{x \in D}.$$

This is an open cover of D and D is compact, so there is a finite subcover of D. In other words, there are $x_1, \ldots, x_n \in D$ such that

$$D \subset \bigcup_{i=1}^{n} \left(x_i - \frac{\delta_{xi}}{2}, x_i + \frac{\delta_{xi}}{2}\right).$$

Let

$$\delta = \min \left\{ \frac{\delta_{x_i}}{2} : i = 1, \ldots, n \right\}.$$

Now suppose $x, y \in D$ and $|x - y| < \delta$. There is an integer i among $1, \ldots, n$ such that

$$x \in \left(x_i - \frac{\delta_{x_i}}{2}, x_i + \frac{\delta_{x_i}}{2} \right).$$

Now $|x - y| < \delta < \frac{\delta_{x_i}}{2}$, making

$$|y - x_i| \leq |y - x| + |x - x_i| < \frac{\delta_{x_i}}{2} + \frac{\delta_{x_i}}{2} = \delta_{x_i}.$$

Hence,

$$|f(x) - f(y)| \leq |f(x) - f(x_i)| + |f(x_i) - f(y)| < \frac{\epsilon}{2} + \frac{\epsilon}{2} = \epsilon.$$

Therefore, f is uniformly continuous.

In a broad sense, this theorem has a converse. Suppose $f: D \to R$ is uniformly continuous. As has been pointed out, f has a limit at each accumulation point of D. As in Exercise 26, let D' be the set of accumulation points of D and $\overline{D} = D \cup D'$. Define $g: \overline{D} \to R$ by $g(x) = f(x)$ for all $x \in D$; and if $x_0 \in \overline{D} \setminus D$, define $g(x_0) = \lim_{x \to x_0} f(x)$. The function $g: \overline{D} \to R$ is called an *extension* of $f: D \to R$ and indeed turns out to be a continuous function. The proof of this will be a challenging exercise for the reader. Thus, if $f: D \to R$ is uniformly continuous with D bounded, then f has a continuous extension $g: \overline{D} \to$ R where $\overline{D}$ is closed and bounded and, hence, compact.

3.4 PROPERTIES OF CONTINUOUS FUNCTIONS

Some of the reasons for considering continuous functions are the special properties they possess. In the next few pages we consider some of these properties and their applications.

3.9 THEOREM If $f: E \to R$ is uniformly continuous and E is a bounded set, then $f(E)$ is a bounded set.

Proof. Choose $\epsilon = 1$. There is $\delta > 0$ such that, if $|x - y| < \delta$, then $|f(x) - f(y)| < 1$. Since E is a bounded set, there are $x_1, x_2, \ldots, x_n$ in E such that

$$E \subset \bigcup_{i=1}^{n} (x_i - \delta, x_i + \delta),$$

but then

$$f(E) \subset \bigcup_{i=1}^{n} (f(x_i) - 1, f(x_i) + 1),$$

since $|x - x_i| < \delta$, $x \in E$, implies that $|f(x) - f(x_i)| < 1$. Thus $f(E)$ is bounded.

Suppose $f\colon E \to R$ is continuous with E compact. Then f is uniformly continuous and E is bounded, so $f(E)$ is bounded. If we would prove that $f(E)$ was closed, then $f(E)$ would be compact and we could say that the continuous image of a compact set is compact. Before proceeding, we need the following lemma.

LEMMA Let E be compact and $\{x_n\}_{n=1}^{\infty}$ be a sequence of points in E. Then there is x_0 in E and a subsequence $\{x_{nk}\}_{n=1}^{\infty}$ that converges to x_0.

Proof. Since E is compact, it is both closed and bounded. Thus $\{x_n\}_{n=1}^{\infty}$, being bounded, has a convergent subsequence $\{x_{nk}\}_{n=1}^{\infty}$. The limit of this subsequence x_0 belongs to E or is an accumulation point of E. (See Theorem 1.17.) But E is closed, hence $x_0 \in E$.

3.10 THEOREM Let $f\colon E \to R$ be continuous with E compact. Then $f(E)$ is compact.

Proof. As pointed out in the preceding paragraph, we need only show that $f(E)$ is closed since we already know $f(E)$ is bounded. Let y_0 be an accumulation point of $f(E)$. Then there is a sequence $\{y_n\}_{n=1}^{\infty}$ of points in $f(E)$ such that $y_n \neq y_0$ for all n and $\{y_n\}_{n=1}^{\infty}$ converges to y_0 (see Theorem 1.17.) Since $y_n \in f(E)$ for each n, there is a sequence $\{x_n\}_{n=1}^{\infty}$ of points in E such that $f(x_n) = y_n$ for each n. There is a subsequence $\{x_{nk}\}_{k=1}^{\infty}$ that converges—call the limit x_0—and $x_0 \in E$ by the lemma. The sequence $\{f(x_{nk})\}_{k=1}^{\infty}$ converges to $f(x_0)$ since f is continuous at x_0. But

$$\{f(x_{nk})\}_{k=1}^{\infty} = \{(y_{nk})\}_{k=1}^{\infty}$$

is a subsequence of $\{y_n\}_{n=1}^{\infty}$ and converges to y_0, so $y_0 = f(x_0) \in f(E)$. Therefore, $f(E)$ is closed.

The following important corollary follows from Theorem 3.10.

3.11 COROLLARY If $f\colon E \to R$ is continuous with E compact, then there are x_1, x_2 in E such that, for all $x \in E$,

$$f(x_1) \leq f(x) \leq f(x_2).$$

Proof. By Theorem 3.10, $f(E)$ is a compact set, and then, by Exercise 33, both $\sup f(E)$ and $\inf f(E)$ belong to $f(E)$. Thus there are x_1 and x_2 in E such that

$$f(x_1) = \inf f(E) \quad \text{and} \quad f(x_2) = \sup f(E).$$

Therefore, for all x in E,

$$f(x_1) \leqslant f(x) \leqslant f(x_2).$$

In essence, this theorem tells us that if $f: E \to R$ is continuous with E compact, then not only is f a bounded function but f actually assumes a maximum and a minimum value on the set E.

Let us explore some examples to illustrate this theorem.

■ **Example 3.7** Consider $f: (0, 1) \to R$ defined by $f(x) = \frac{1}{x}$. The function is continuous but not bounded from above. It is bounded from below, but $\inf f((0, 1)) = 1$, and there is no $x \in (0, 1)$ such that $f(x) = 1$. ■

■ **Example 3.8** The function $g: (0, 1) \to R$ defined by $g(x) = x$ is certainly bounded, but $\inf g((0, 1)) = 0$ and $\sup g((0, 1)) = 1$, neither of which is a value of g for the reason that $(0, 1)$ is not compact. ■

To lay the groundwork for a new theorem worthy of note, suppose $f: E \to R$ is 1–1. Then there is a unique function $f^{-1}: f(E) \to E$ such that $f \circ f^{-1}(x) = x$ for each $x \in f(E)$ and $f^{-1} \circ f(y) = y$ for each $y \in E$. A natural question to raise is the following: If f is continuous, is f^{-1} continuous? The next theorem gives a partial answer.

3.12 THEOREM Suppose $f: E \to R$ is continuous and 1–1 with E compact. Then $f^{-1}: f(E) \to E$ is continuous.

Proof. Let $\{y_n\}_{n=1}^{\infty}$ be any sequence in $f(E)$ converging to $y_0 \in f(E)$. We will show that $\{f^{-1}(y_n)\}_{n=1}^{\infty}$ converges to $f^{-1}(y_0)$. For ease of understanding, we will write $x_n = f^{-1}(y_n)$ for each n. Thus $f(x_n) = y_n$. Since E is compact, the sequence $\{x_n\}_{n=1}^{\infty}$ is bounded and Theorem 1.15 applies. Let $\{x_{nk}\}_{k=1}^{\infty}$ be any convergent subsequence of $\{x_n\}_{n=1}^{\infty}$; call the limit z_0. Now z_0 belongs to E since E is closed, and so f is continuous at z_0; thus, $\{f(x_{nk})\}_{k=1}^{\infty}$ converges to $f(z_0)$. But $\{f(x_{nk})\}_{k=1}^{\infty} = \{y_{nk}\}_{k=1}^{\infty}$ is a subsequence of $\{y_n\}_{n=1}^{\infty}$, hence it converges to $y_0 = f(x_0)$. Since f is 1–1 and $f(x_0) = f(z_0)$, $x_0 = z_0$. By Theorem 1.15, $\{x_n\}_{n=1}^{\infty}$ converges to x_0; or, in other words, $\{f^{-1}(y_n)\}_{n=1}^{\infty}$ converges to $f^{-1}(y_0)$. Thus f^{-1} is continuous at y_0.

■ **Example 3.9** To show the need for the hypothesis that E be compact, we shall exhibit an example of a function $f: E \to R$ that is 1–1 but such that $f^{-1}: f(E) \to E$ is not continuous. Of course, E must be noncompact. Let

$$E = [0, 1] \cup [2, 3)$$

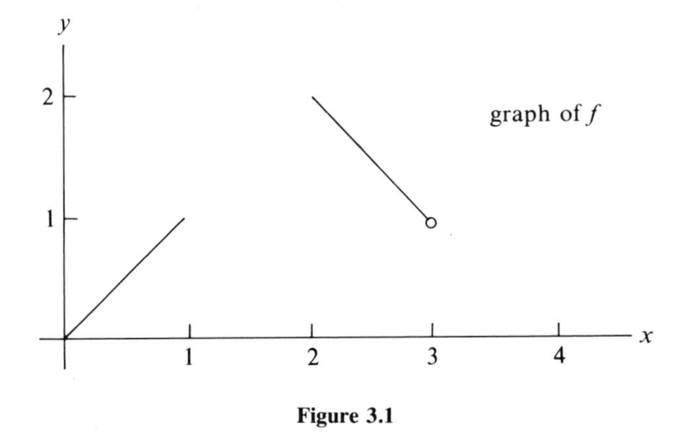

Figure 3.1

and define $f(x) = x$ for $0 \leq x \leq 1$ and $f(x) = 4 - x$ for $2 \leq x < 3$. It is left to the readers to satisfy themselves that f is indeed 1–1 and continuous. See Figure 3.1 for the graph of f. Now $f(E) = [0, 2]$ and for $0 \leq x \leq 1$, $f^{-1}(x) = x$ while for $1 < x \leq 2$, $f^{-1}(x) = 4 - x$. See Figure 3.2 for the graph of f^{-1}. Now consider the sequence

$$\left\{1 + \frac{(-1)^n}{n}\right\}_{n=1}^{\infty},$$

which converges to 1. For n odd,

$$f^{-1}\left(1 + \frac{(-1)^n}{n}\right) = 1 + \frac{(-1)^n}{n},$$

and for n even,

$$f^{-1}\left(1 + \frac{(-1)^n}{n}\right) = 4 - \left(1 + \frac{(-1)^n}{n}\right) = 3 - \frac{(-1)^n}{n}.$$

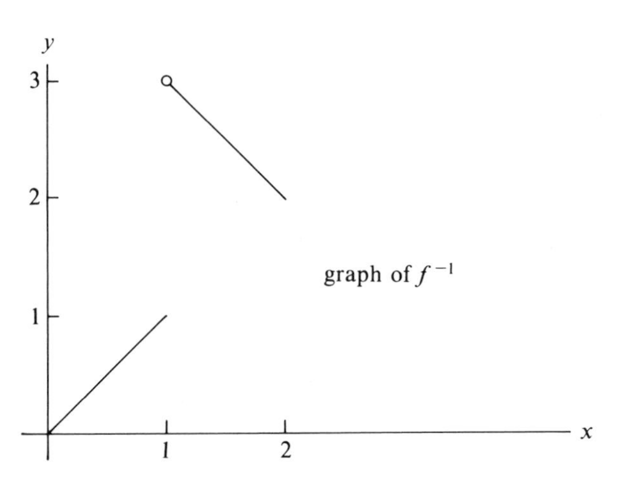

Figure 3.2

It should be clear then that $\left\{f^{-1}\left(1 + \frac{(-1)^n}{n}\right)\right\}_{n=1}^{\infty}$ does not converge, and so f^{-1} fails to be continuous at 1. ■

Consider now a function f: $[-1, 1] \to R$ that is continuous and is such that $f(-1) < 0$ and $f(1) > 0$. An intuitive picture of the geometric nature of the graph of a continuous function leads us to guess that this graph must cross the x-axis somewhere between -1 and $+1$; that is, there is $x \in (-1, 1)$ such that $f(x) = 0$. This indeed turns out to be the case, and in fact it motivates us to state the following theorem.

3.13 BOLZANO INTERMEDIATE-VALUE THEOREM Let f: $[a, b] \to R$ be continuous with $f(a) \leqslant y \leqslant f(b)$ [or $f(b) \leqslant y \leqslant f(a)$]. Then there is $c \in (a, b)$ with $f(c) = y$.

Proof. Let $A = \{x : x \in [a, b] \text{ and } f(x) \leqslant y\}$ and let $B = \{x : x \in [a, b] \text{ and } f(x) \geqslant y\}$. Now, by hypothesis, $a \in A$ and $b \in B$. Let $c = \sup A$. If $f(c) = y$, we are through. Suppose not. Suppose $f(c) < y$. Let

$$\epsilon = \frac{1}{2}[y - f(c)] > 0.$$

By the continuity of f, there is $\delta > 0$ such that $|x - c| < \delta$ and $x \in [a, b]$ implies $|f(x) - f(c)| < \epsilon$; that is,

$$f(x) < f(c) + \epsilon < y.$$

Now since $f(c) < y$, $c \notin B$; hence $c < b$. Thus, there is $x \in [a, b]$ such that $c < x < c + \delta$; hence $f(x) < y$. This, however, is a contradiction to the choice of $c = \sup A$. Thus, $f(c) \nless y$.

Since $f(c) \nless y$, we may suppose $f(c) > y$. In this case, let

$$\epsilon = \frac{1}{2}[f(c) - y] > 0.$$

There is $\delta > 0$ such that if $|x - c| < \delta$ and $x \in [a, b]$, then $|f(x) - f(c)| < \epsilon$; that is

$$f(x) > f(c) - \epsilon > y.$$

Since $f(c) > y$, $c \notin A$; hence $c > a$. Therefore, there is $x_0 \in [a, b]$ with $c - \delta < x_0 < c$. If $x_0 \leqslant x < c$, then $f(x) > y$; hence $x \notin A$. This means that x_0 is also an upper bound for A, contrary to $c = \sup A$.

Since $f(c) > y$ and $f(c) < y$ are impossible, we conclude that $f(c) = y$.

This theorem is, of course, the one that is used in trying to find the zeros of a polynomial. If p is a polynomial with $p(a) < 0$ and $p(b) > 0$, then p has at least one zero between a and b, since every polynomial is continuous. If p is a polynomial of odd degree, then p will have a change of sign, so it necessarily has a zero. Let us prove this fact.

3.14 THEOREM If p is a polynomial of odd degree with real coefficients, then the equation $p(x) = 0$ has at least one real root.

Proof. Assume

$$p(x) = a_0 + a_1x + \cdots + a_nx^n$$

where n is odd and $a_n \neq 0$. If $a_0 = 0$, then $p(0) = a_0 = 0$ and we are through. Assume now that $a_0 \neq 0$, and let

$$\phi(x) = \frac{a_0}{x^n} + \frac{a_1}{x^{n-1}} + \cdots + a_n$$

for all $x \neq 0$. Choose ϵ such that $0 < \epsilon < |a_n|$.
For $i = 0, 1, 2, \ldots, n - 1$, there exists $k_i > 1$ such that

$$\frac{|a_i|}{k_i} < \frac{\epsilon}{n}.$$

Let $K = \max\{k_0, k_1, \ldots, k_{n-1}\}$. Now, if $|x| \geq K$, then

$$\begin{aligned}\left|\frac{a_0}{x^n} + \frac{a_1}{x^{n-1}} + \cdots + \frac{a_{n-1}}{x}\right| &\leq \frac{|a_0|}{|x^n|} + \cdots + \frac{|a_{n-1}|}{|x|} \\ &\leq \frac{|a_0|}{K} + \cdots + \frac{|a_{n-1}|}{K} < \frac{\epsilon}{n} + \cdots + \frac{\epsilon}{n} = \epsilon.\end{aligned}$$

Thus, for $|x| \geq K$, $\phi(x)$ has the same sign as a_n. Now $p(x) = \phi(x)x^n$. So, if $a_n > 0$ with $x \geq K$, then $p(x) > 0$; and, if $x \leq -K$, then $p(x) < 0$. Similarly, if $a_n < 0$ and $x \geq K$, then $p(x) < 0$; and, if $x \leq -K$, then $p(x) > 0$. In any case, we can find points x_1 and x_2 such that $p(x_1) < 0$ and $p(x_2) > 0$; hence, somewhere between x_1 and x_2, p has a zero.

If we combine Corollary 3.11 and Theorem 3.13, we obtain the following result.

3.15 THEOREM If $f\colon [a, b] \to R$ is continuous, there are c and d such that $f([a, b]) = [c, d]$.

Proof. By Corollary 3.11, there are x_1 and x_2 in $[a, b]$ such that $f(x_1) \leq f(x) \leq f(x_2)$ for all $x \in [a, b]$. Let $c = f(x_1)$ and $d = f(x_2)$. By Theorem 3.13, if $c \leq y \leq d$, then there is $x \in [a, b]$ such that $f(x) = y$. Thus, $f([a, b]) = [c, d]$.

■ **Example 3.10** As a further application, we shall show that the equation $x = \cos x$ has at least one solution in the closed interval $[0, \frac{\pi}{2}]$. Consider the function $f(x) = x - \cos x$ on $[0, \frac{\pi}{2}]$. Clearly, f is continuous, $f(0) = -1$, and $f(\frac{\pi}{2}) = \frac{\pi}{2}$. Therefore, there is $x \in [0, \frac{\pi}{2}]$ such that $f(x) = 0$; in other words, $x = \cos x$. ■

3.16 THEOREM Let $f\colon [a, b] \to R$ be continuous and 1–1. Then f is monotone.

Proof. Suppose f is 1–1 and not monotone. Then there are x, y, $z \in [a, b]$ such that $x < y < z$, and $f(x) < f(y)$ and $f(z) < f(y)$ or $f(x) > f(y)$ and $f(z) > f(y)$. Let us suppose the latter; in fact, let us suppose $f(x) > f(z) > f(y)$. The other cases may be handled in a similar fashion. By the Intermediate-Value Theorem, there is $w \in (x, y)$ such that $f(w) = f(z)$, contrary to f being 1–1.

Let us now return to a topic considered at the end of Chapter 2. Let $f\colon [\alpha, \beta] \to R$ be monotone. Then $D = \{x : x \in (\alpha, \beta)$ and f does not have a limit at $x\}$ is countable; and if f has a limit at $x_0 \in (\alpha, \beta)$, then $\lim_{x \to x_0} f(x) = f(x_0)$. The discussion of the behavior of f at α and at β were left to the reader. It is no betrayal of confidence to state now that f has a limit at α and at β and that

$$\lim_{x \to \alpha} f(x) = \inf \{f(x) : \alpha < x \leq \beta\}$$

and

$$\lim_{x \to \beta} f(x) = \sup \{f(x) : \alpha \leq x < \beta\}$$

if f is increasing. In the case where f is decreasing, both limits still exist, but

$$\lim_{x \to \alpha} f(x) = \sup \{f(x) : \alpha < x \leq \beta\}$$

and

$$\lim_{x \to \beta} f(x) = \inf \{f(x) : \alpha \leq x < \beta\}$$

Now it is clear that f is continuous at x for each $x \in (\alpha, \beta) \setminus D$, and possibly at α and β depending on the definition of f at α and β. In particular, the set of points at which f is discontinuous is countable.

In Chapter 0, we stated that not all sets were countable; indeed, we stated that R was uncountable and promised a proof of that fact later. Well, later is now! The next theorem states that $[0, 1]$ is uncountable, and we leave as Exercise 42 the proof that R is equivalent to $[0, 1]$.

3.17 THEOREM The set $[0, 1]$ is uncountable.

Proof. Let us assume the theorem is false and $[0, 1]$ is countable. Then there is a 1–1 function $T: J \to [0, 1]$ such that the image of T is $[0, 1]$. Now $T(1) \in [0, 1]$, and so there is a closed interval $[a_1, b_1] \subset [0, 1]$ such that $a_1 < b_1$ and $T(1) \not\in [a_1, b_1]$. There is a closed interval $[a_2, b_2] \subset [a_1, b_1]$ such that $a_2 < b_2$ and $T(2) \not\in [a_2, b_2]$. Continuing in this fashion, we construct a sequence $\{[a_n, b_n]\}_{n=1}^{\infty}$ of intervals such that for each n,

1. $[a_n, b_n] \subset [a_{n-1}, b_{n-1}] \subset [a_{n-2}, b_{n-2}] \subset \cdots \subset [a_1, b_1] \subset [0, 1]$.
2. $T(n) \not\in [a_n, b_n]$.
3. $a_n < b_n$.

Since T maps J onto $[0, 1]$ and $T(n) \not\in [a_n, b_n]$, the set $\cap_{n=1}^{\infty} [a_n, b_n]$ is empty. For each n, define $A_n = R \setminus [a_n, b_n]$. Now

$$\bigcup_{n=1}^{\infty} A_n = \bigcup_{n=1}^{\infty} (R \setminus [a_n, b_n]) = R \setminus (\bigcap_{n=1}^{\infty} [a_n, b_n]) = R;$$

hence $\{A_n\}_{n=1}^{\infty}$ is an open cover of $[0, 1]$. By Theorem 3.7, $[0, 1]$ is compact, so there are integers $n_1, n_2, \ldots, n_r$ such that

$$[0, 1] \subset \bigcup_{i=1}^{r} A_{n_i} = A_m,$$

where $n = \max\{n_1, n_2, \ldots, n_r\}$. This means that

$$[0, 1] \subset A_m = R \setminus [a_m, b_m],$$

contrary to the fact that $[a_m, b_m] \subset [0, 1]$.

Thus, the assumption that $[0, 1]$ is countable leads us to a contradiction; so the assumption is false. Therefore, the set $[0, 1]$ is not a countable set.

EXERCISES

3.1 CONTINUITY OF A FUNCTION AT A POINT

1. Define $f: R \to R$ by $f(x) = 3x^2 - 2x + 1$. Show that f is continuous at 2.

2. Define $f: [-4, 0] \to R$ by $f(x) = \dfrac{2x^2 - 18}{x + 3}$ for $x \neq -3$ and $f(-3) = -12$. Show that f is continuous at -3.

3. Use Theorem 3.1 to prove that $\{\sqrt[n]{e^{n+1}}\}_{n=1}^{\infty}$ is convergent and find the limit. You may assume that the function $f(x) = e^x$ is continuous on R.

4. If $x_0 \in E$, x_0 is not an accumulation point of E, and $f: E \to R$, prove that, for every sequence $\{x_n\}_{n=1}^{\infty}$ converging to x_0 with $x_n \in E$ for all n, $\{f(x_n)\}_{n=1}^{\infty}$ converges to $f(x_0)$.

5. Define $f: (0, 1) \to R$ by $f(x) = \dfrac{1}{\sqrt{x}} - \sqrt{\dfrac{x + 1}{x}}$. Can one define $f(0)$ to make f continuous at 0? Explain.

6. Prove that $f(x) = \sqrt{x}$ is continuous for all $x \geq 0$.

7. Suppose f: $(a, b) \to R$ is continuous and $f(r) = 0$ for each rational number $r \in (a, b)$. Prove that $f(x) = 0$ for all $x \in (a, b)$.

8. Define f: $(0, 1) \to R$ by $f(x) = x \sin \frac{1}{x}$. Can one define $f(0)$ to make f continuous at 0? Explain.

***9.** Suppose f: $E \to R$ is continuous at x_0 and $x_0 \in F \subset E$. Define g: $F \to R$ by $g(x) = f(x)$ for all $x \in F$. Prove that g is continuous at x_0. Show by example that the continuity of g at x_0 need not imply the continuity of f at x_0.

10. Define f: $R \to R$ by $f(x) = 8x$ if x is rational and $f(x) = 2x^2 + 8$ if x is irrational. Prove from the definition of continuity that f is continuous at 2 and discontinuous at 1.

3.2 ALGEBRA OF CONTINUOUS FUNCTIONS

***11.** Let p and q be polynomials and x_0 be a zero of q of multiplicity m. Prove that $\frac{p}{q}$ can be assigned a value at x_0 such that the function thus defined will be continuous there iff x_0 is a zero of p of multiplicity greater than or equal to m.

12. Let f: $D \to R$ be continuous at $x_0 \in D$. Prove that there is $M > 0$ and a neighborhood Q of x_0 such that $|f(x)| \leq M$ for all $x \in Q \cap D$.

13. If f: $D \to R$ is continuous at $x_0 \in D$, prove that the function $|f|$: $D \to R$ such that $|f|(x) = |f(x)|$ is continuous at x_0.

14. Suppose f, g: $D \to R$ are both continuous on D. Define h: $D \to R$ by $h(x) = \max(f(x), g(x))$. Show that h is continuous on D.

15. Assume the continuity of $f(x) = e^x$ and $g(x) = \ln x$. Define $h(x) = x^x$ by $x^x = e^{x \ln x}$. Show h is continuous for $x > 0$.

16. Suppose f: $D \to R$ with $f(x) \geq 0$ for all $x \in D$. Show that, if f is continuous at x_0, then $\sqrt{f}$ is continuous at x_0.

17. Define f: $R \to R$ as follows:

$$f(x) = x - [x] \text{ if } [x] \text{ is even,}$$
$$f(x) = x - [x + 1] \text{ if } [x] \text{ is odd.}$$

Determine those points where f is continuous. Justify.

3.3 UNIFORM CONTINUITY: OPEN, CLOSED, AND COMPACT SETS

18. Let f, g: $D \to R$ be uniformly continuous. Prove that $f + g$: $D \to R$ is uniformly continuous. What can be said about fg? Justify.

19. Let f: $A \to B$ and g: $B \to C$ be uniformly continuous. What can be said about $g \circ f$: $A \to C$? Justify.

20. Define f: $[3.4, 5] \to R$ by $f(x) = \dfrac{2}{x - 3}$. Show that f is uniformly continuous on $[3.4, 5]$ without using Theorem 3.8—that is, use the methods of Example 3.2.

21. Define f: $(2, 7) \to R$ by $f(x) = x^3 - x + 1$. Show that f is uniformly continuous on $(2, 7)$ without using Theorem 3.8—that is, use the methods of Example 3.3.

22. Define f: $(0, 1) \to R$ by $f(x) = \frac{1}{x}$. Show that f is not uniformly continuous.

23. A function f: $R \to R$ is periodic iff there is a real number $h \neq 0$ such that $f(x + h) = f(x)$ for all $x \in R$. Prove that if f: $R \to R$ is periodic and continuous, then f is uniformly continuous.

***24.** Let $E \subset R$. Prove that E is closed if, for every x_0 such that there is a sequence $\{x_n\}_{n=1}^{\infty}$ of points of E converging to x_0, it is true that $x_0 \in E$. In other words, prove E is closed if it contains all limits of sequences of members of E.

***25.** Prove that every set of the form $\{x : a < x < b\}$ is open and every set of the form $\{x : a \leq x \leq b\}$ is closed.

26. Let $D \subset R$ and let D' be the set of accumulation points of D. Prove that $\overline{D} = D \cup D'$ is closed and that if F is any closed set that contains D, then $\overline{D}' \subset F$. $\overline{D}$ is called the *closure* of D.

27. If $D \subset R$ is bounded, prove that $\overline{D}$ is bounded.

28. Suppose $f: R \to R$ is continuous and let $r_0 \in R$. Prove that $\{x : f(x) \neq r_0\}$ is an open set.

29. Suppose $f: [a, b] \to R$ and $g: [a, b] \to R$ are both continuous. Let $T = \{x : f(x) = g(x)\}$. Prove that T is closed.

30. If $D \subset R$, then $x \in D$ is said to the the *interior point* of D iff there is a neighborhood Q of x such that $Q \subset D$. Define D° to be the set of interior points of D. Prove that D° is open and that if S is any open set contained in D, then $S \subset D^\circ$. D° is called the *interior* of D.

31. Find an open cover of $\{x : x > 0\}$ with no finite subcover.

32. Find an open cover of $(1, 2)$ with no finite subcover.

***33.** Let E be compact and nonempty. Prove that E is bounded and that $\sup E$ and $\inf E$ both belong to E.

34. If $E_1, \ldots, E_n$ are compact, prove that $E = \cup_{i=1}^{n} E_i$ is compact.

35. Let $f: [a, b] \to R$ have a limit at each $x \in [a, b]$. Prove that f is bounded.

36. Suppose $f: D \to R$ is continuous with D compact. Prove that $\{x : 0 \leq f(x) \leq 1\}$ is compact.

37. Suppose that $f: R \to R$ is continuous and has the property that for each $\epsilon > 0$, there is $M > 0$ such that if $|x| \geq M$, then $|f(x)| < \epsilon$. Show that f is uniformly continuous.

38. Give an example of a function $f: R \to R$ that is continuous and bounded but not uniformly continuous.

3.4 PROPERTIES OF CONTINUOUS FUNCTIONS

39. Find an interval of length 1 that contains a root of the equation $xe^x = 1$.

40. Find an interval of length 1 that contains a root of the equation $x^3 - 6x^2 + 2.826 = 0$.

41. Suppose $f: [a, b] \to R$ is continuous and $f(b) \leq y \leq f(a)$. Prove that there is $c \in [a, b]$ such that $f(c) = y$.

42. Prove that R is equivalent to $[0, 1]$.

43. Suppose that $f: [a, b] \to [a, b]$ is continuous. Prove that there is at least one fixed point in $[a, b]$—that is, x such that $f(x) = x$.

44. If $f: [a, b] \to R$ is 1–1 and has the intermediate-value property—that is, if y is between $f(u)$ and $f(v)$, there is x between u and v such that $f(x) = y$—show that f is continuous. (*Hint:* First show that f is monotone.)

45. Prove that there is no continuous function $f: R \to R$ such that, for each $c \in R$, the equation $f(x) = c$ has exactly two solutions.

MISCELLANEOUS

46. Let $f: R \to R$ be additive. (See the project at the end of Chapter 2.) That is, $f(x + y) = f(x) + f(y)$ for all $x, y \in R$. In addition, assume that there are $M > 0$ and $a > 0$ such that if $x \in [-a, a]$, then $|f(x)| \leq M$. Prove that f is uniformly continuous. In particular, prove that there is a real number m such that $f(x) = mx$ for all $x \in R$.

47. Let $f: [a, b] \to R$ be continuous, and define $g: [a, b] \to R$ by

$$g(t) = \sup \{f(x) : a \leq x \leq t\}.$$

Prove that g is continuous. (This exercise may be difficult if you didn't cover Section 2.4. In fact, you might also look at Exercise 25 in Chapter 2.)

48. Suppose that $g: D \to R$ is continuous at x_0 and that x_0 is also an accumulation point of D. Define $D_0 = \{x : g(x) \neq 0\}$. If $g(x_0) \neq 0$, prove that x_0 is an accumulation point of D_0.

49. Suppose $f: D \to R$, $g: E \to R$ and $x_0 \in D \cap E$. Suppose further that there is $\epsilon > 0$ such that $D \cap [x_0 - \epsilon, x_0 + \epsilon] = E \cap [x_0 - \epsilon, x_0 + \epsilon]$ and $f(x) = g(x)$ for all $x \in D \cap E \cap [x_0 - \epsilon, x_0 + \epsilon]$. Prove that f is continuous at x_0 iff g is continuous at x_0.

PROJECT

The purpose of this project is to utilize continuity of a certain function to establish the convergence of an iterative process for approximating the square root of 2. The process is a familiar one—the divide-and-average process.

We wish to define a sequence recursively as follows:

$$a_1 = 2 \quad \text{and} \quad a_n = \frac{1}{2}\left(a_{n-1} + \frac{2}{a_{n-1}}\right) \quad \text{for } n \geq 2$$

Our hypothesis is that $\{a_n\}_{n=1}^{\infty}$ converges to $\sqrt{2}$. To show this, we need some information about the function $f(x) = \frac{1}{2}(x + \frac{2}{x})$, $x > 0$.

1. Show that for all $x > 0$, $f(x) \geq \sqrt{2}$. This can be a bit tricky. Use some algebra to write $(x + \frac{2}{x})^2 - 8$ as a square to establish $(x + \frac{2}{x})^2 \geq 8$ and, hence, $f(x) \geq \sqrt{2}$.
2. Show that for $x \geq \sqrt{2}$, $f(x) \leq x$.
3. Define a sequence as follows: $a_1 = 2$ and $a_n = f(a_{n-1})$ for $n \geq 2$. Prove that $\{a_n\}_{n=1}^{\infty}$ converges. (*Hint:* Show that the sequence is decreasing and bounded from below.)
4. Let x_0 be the limit of the sequence defined in part 3. Prove that $x_0 = \sqrt{2}$. Use the continuity of f.

CHAPTER 4

Differentiation

Consider the function $f(x) = x^3$ and choose a fixed point x_0. Choose another point x, and through the points $(x, f(x))$ and $(x_0, f(x_0))$, draw a line $L(x)$. The slope of this line is

$$\frac{f(x) - f(x_0)}{x - x_0},$$

and an easy computation shows us that

$$\frac{f(x) - f(x_0)}{x - x_0} = \frac{x^3 - x_0^3}{x - x_0} = \frac{(x - x_0)(x^2 + xx_0 + x_0^2)}{x - x_0}$$
$$= x^2 + xx_0 + x_0^2.$$

Upon careful examination of this expression for the slope of the line $L(x)$, it becomes clear that the slope of $L(x)$ has a limit at x_0 and that limit is $3x_0^2$. If there is a reasonable definition of the slope of a curve at a point, then it should arise from the approach just used.

Suppose now that the object is moving in a rectilinear path with the distance from a fixed point s_0 given at time t by $f(t)$. For fixed choice of t_0 and $t \neq t_0$, the ratio

$$\frac{f(t) - f(t_0)}{t - t_0}$$

gives the average value of the velocity over the time interval $[t, t_0]$ or $[t_0, t]$ as $t < t_0$ or $t > t_0$. Thus, for each $t \neq t_0$, it is possible to compute an average velocity, $A(t)$. If A has a limit at t_0, it seems reasonable to call this limit the *velocity* at t_0.

The two preceding paragraphs are given as a reminder of the many reasons for considering the notion of the derivative of a function at a point. Our approach, though conventional, should extend the reader's knowledge of and feeling for differentiation.

4.1 THE DERIVATIVE OF A FUNCTION

DEFINITION Let $f\colon D \to R$ with x_0 an accumulation point of D and $x_0 \in D$. For each $x \in D$ with $x \neq x_0$, define

$$T(x) = \frac{f(x) - f(x_0)}{x - x_0}.$$

The function f is said to be *differentiable at* x_0 (or *has a derivative at* x_0) iff T has a limit at x_0, and we write $\lim_{x \to x_0} T(x) = f'(x_0)$. The number $f'(x_0)$ is called the *derivative* of f at x_0. If f is differentiable for each $x \in E \subset D$, we say f is differentiable on E.

First we shall state a basic theorem that gives a condition for differentiability in terms of sequences. Since the definition of differentiability is based on the existence of the limit of a function, this theorem should come as no surprise.

4.1 THEOREM Suppose $f\colon D \to R$, $x_0 \in D$, and x_0 is an accumulation point of D. Then f is differentiable at x_0 iff for every sequence $\{x_n\}_{n=1}^{\infty}$ of points of $D \setminus \{x_0\}$ converging to x_0, the sequence

$$\left\{\frac{f(x_n) - f(x_0)}{x_n - x_0}\right\}_{n=1}^{\infty}$$

converges.

No formal proof will be supplied here, since it should be clear from Theorem 2.1 that the condition concerning sequences is equivalent to the condition that the function T given in the definition of differentiability has a limit at x_0.

Note also that if f is differentiable at x_0 and $\{x_n\}_{n=1}^{\infty}$ is a sequence of points of $D \setminus \{x_0\}$ converging to x_0, then

$$\left\{\frac{f(x_n) - f(x_0)}{x_n - x_0}\right\}_{n=1}^{\infty}$$

converges to $f'(x_0)$.

The next few examples will further one's understanding of the definition of differentiation.

■ **Example 4.1** Consider the function $f(x) = |x|$ for all $x \in R$. Set $x_0 = 0$, and consider the sequence $\left\{\frac{(-1)^n}{n}\right\}_{n=1}^{\infty}$; this sequence converges to zero, but zero is not a term of this sequence. For n even,

$$\frac{(-1)^n}{n} = \frac{1}{n} \quad \text{and} \quad \frac{f\left(\frac{1}{n}\right) - f(0)}{\frac{1}{n}} = 1;$$

for n odd,

$$\frac{(-1)^n}{n} = -\frac{1}{n} \text{ and } \frac{f\left(-\frac{1}{n}\right) - f(0)}{-\frac{1}{n}} = -1,$$

so the sequence

$$\left\{ \frac{f\left(\frac{(-1)^n}{n}\right) - f(0)}{\frac{(-1)^n}{n}} \right\}_{n=1}^{\infty}$$

does not converge. Thus, by Theorem 4.1, f is not differentiable at zero. The reader should sketch the graph of $|x|$ to see why this happens. The graph has a "sharp" corner at zero with the slope suddenly changing from -1 to $+1$ as x increases from negative to positive values, and of course the slope is not defined at zero. ■

■ **Example 4.2** For the moment, assume the usual facts about differentiating the product, sum, and composition of the elementary functions. Define $f(x) = x \sin \frac{1}{x}$ for $x \neq 0$ and $f(0) = 0$. The fact that f is differentiable at any point other than zero follows from the assumptions made at the beginning of this paragraph. For $x \neq 0$,

$$\frac{f(x) - f(0)}{x - 0} = \sin \frac{1}{x}$$

does not have a limit at zero; hence f is not differentiable at zero, although it is continuous there. ■

■ **Example 4.3** In an attempt to find a "nicer" function, define $f(x) = x^2 \sin \frac{1}{x}$ for $x \neq 0$ and $f(0) = 0$. Again, assume the basic facts about differentiation of products and composition of elementary functions. Now, f has a derivative at x if $x \neq 0$, and $f'(x) = 2x \sin \frac{1}{x} - \cos \frac{1}{x}$. So $f'(x)$ is defined for $x \neq 0$, but f' does not have a limit at zero. One might suspect that this means that f is not differentiable at 0; however, this is false. For $x \neq 0$,

$$\frac{f(x) - f(0)}{x - 0} = x \sin \frac{1}{x},$$

which has a limit at zero, namely, $f'(0) = 0$. In this case we have a function with a derivative at every point, but the derivative is not continuous at zero; in fact, it does not have a limit there. ■

■ **Example 4.4** Finally consider $f(x) = x^3 \sin \frac{1}{x}$ for $x \neq 0$ and $f(0) = 0$. Again, using the usual rules (to be proven in the next section), f has a derivative everywhere and f' is continuous everywhere but fails to be differentiable at one point. Can you guess the point? See Exercise 3. ■

Let us now return to the initial definition and seek the facts pertinent to functions differentiable on a set. First of all, consider the difference quotient

$$\frac{f(x) - f(x_0)}{x - x_0} = T(x).$$

For x close to x_0, the denominator is close to zero; hence, in order that T have a limit at x_0, the numerator must be close to zero—otherwise T would not be bounded near x_0. We are led to predict that, if f is differentiable at x_0, then f must be continuous at x_0. The proof may be accomplished by observing that

$$f(x) - f(x_0) = \frac{f(x) - f(x_0)}{x - x_0}(x - x_0)$$

for $x \neq x_0$.

4.2 THEOREM Let $f\colon D \to R$ be differentiable at x_0. (It is tacitly assumed that $x_0 \in D$ and that x_0 is an accumulation point of D.) Then f is continuous at x_0.

Proof. Let $T\colon D \setminus \{x_0\} \to R$ be defined by

$$T(x) = \frac{f(x) - f(x_0)}{x - x_0}.$$

T has a limit at x_0 and $\lim_{x \to x_0} T(x) = f'(x_0)$. For $x \neq x_0$,

$$\begin{aligned} f(x) &= \frac{f(x) - f(x_0)}{x - x_0}(x - x_0) + f(x_0) \\ &= T(x)(x - x_0) + f(x_0). \end{aligned}$$

Now f is a sum of two functions, one of which is constant and the other the product of two functions, each of which has a limit at x_0; so f has a limit at x_0 and

$$\begin{aligned} \lim_{x \to x_0} f(x) &= \lim_{x \to x_0} [T(x)(x - x_0) + f(x_0)] \\ &= [\lim_{x \to x_0} T(x)][\lim_{x \to x_0} (x - x_0)] + \lim_{x \to x_0} f(x_0) \\ &= f'(x_0) \cdot 0 + f(x_0) = f(x_0). \end{aligned}$$

Thus, f is continuous at x_0.

4.2 THE ALGEBRA OF DERIVATIVES

The next theorem presents the usual facts from calculus about the sum, product, and quotient of differentiable functions.

4.3 THEOREM Suppose f, g: $D \to R$ are differentiable at x_0. Then

a. $f + g$ is differentiable at x_0 and

$$(f + g)'(x_0) = f'(x_0) + g'(x_0).$$

b. fg is differentiable at x_0 and

$$(fg)'(x_0) = f(x_0)g'(x_0) + f'(x_0)g(x_0).$$

c. If $g(x_0) \neq 0$, then $\frac{f}{g}$ (the domain is the set of all x such that $g(x) \neq 0$) is differentiable at x_0 and

$$\left(\frac{f}{g}\right)'(x_0) = \frac{f'(x_0)g(x_0) - g'(x_0)f(x_0)}{[g(x_0)]^2}.$$

Proof. We shall appeal to Theorem 4.1 to prove this theorem. Let $\{x_n\}_{n=1}^{\infty}$ be any sequence of points in $D \setminus \{x_0\}$ converging to x_0. Since f and g are differentiable at x_0.

$$\left\{\frac{f(x_n) - f(x_0)}{x_n - x_0}\right\}_{n=1}^{\infty}$$

converges to $f'(x_0)$ and

$$\left\{\frac{g(x_n) - g(x_0)}{x_n - x_0}\right\}_{n=1}^{\infty}$$

converges to $g'(x_0)$. Thus

$$\left\{\frac{(f + g)(x_n) - (f + g)(x_0)}{x_n - x_0}\right\}_{n=1}^{\infty} = \left\{\frac{f(x_n) - f(x_0)}{x_n - x_0} + \frac{g(x_n) - g(x_0)}{x_n - x_0}\right\}_{n=1}^{\infty}$$

converges to $f'(x_0) + g'(x_0)$. This means that $f + g$ is differentiable at x_0 and that

$$(f + g)'(x_0) = f'(x_0) + g'(x_0).$$

By Theorem 4.2, f is continuous at x_0, and so $\{f(x_n)\}_{n=1}^{\infty}$ converges to $f(x_0)$. Thus

$$\left\{\frac{(fg)(x_n) - (fg)(x_0)}{x_n - x_0}\right\}_{n=1}^{\infty}$$
$$= \left\{\frac{f(x_n)g(x_n) - f(x_n)g(x_0) + f(x_n)g(x_0) - f(x_0)g(x_0)}{x_n - x_0}\right\}_{n=1}^{\infty}$$

$$= \left\{ f(x_n)\frac{g(x_n) - g(x_0)}{x_n - x_0} + g(x_0)\frac{f(x_n) - f(x_0)}{x_n - x_0} \right\}_{n=1}^{\infty}$$

converges to $f(x_0)g'(x_0) + g(x_0)f'(x_0)$. Thus, fg is differentiable at x_0 and

$$(fg)'(x_0) = f(x_0)g'(x_0) + g(x_0)f'(x_0).$$

Let $D' = \{x : x \in D \text{ and } g(x) \neq 0\}$. Then D' is the domain of $\frac{f}{g}$, and x_0 is an accumulation point of D' by the continuity of g at x_0. Let $\{x_n\}_{n=1}^{\infty}$ be any sequence of points of $D' \setminus \{x_0\}$ converging to x_0. Then, since g is continuous at x_0 and $g(x_0) \neq 0$, $\frac{1}{g}$ is continuous at x_0; hence, $\{\frac{1}{g}(x_n)\}_{n=1}^{\infty}$ converges to $\frac{1}{g}(x_0)$. Now

$$\left\{ \frac{\left(\frac{f}{g}\right)(x_n) - \left(\frac{f}{g}\right)(x_0)}{x_n - x_0} \right\}_{n=1}^{\infty}$$

$$= \left\{ \frac{f(x_n)g(x_0) - g(x_n)f(x_0)}{g(x_n)g(x_0)(x_n - x_0)} \right\}_{n=1}^{\infty}$$

$$= \left\{ \frac{1}{g(x_n)g(x_0)} \times \left[\frac{f(x_n)g(x_0) - f(x_0)g(x_0) + f(x_0)g(x_0) - g(x_n)f(x_0)}{x_n - x_0} \right] \right\}_{n=1}^{\infty}$$

$$= \left\{ \frac{1}{g(x_n)g(x_0)} \left[g(x_0)\frac{f(x_n) - f(x_0)}{x_n - x_0} - f(x_0)\frac{g(x_n) - g(x_0)}{x_n - x_0} \right] \right\}_{n=1}^{\infty}$$

converges to

$$\frac{1}{[g(x_0)]^2}[g(x_0)f'(x_0) - f(x_0)g'(x_0)].$$

Thus, $\frac{f}{g}$ is differentiable at x_0 and

$$\left(\frac{f}{g}\right)'(x_0) = \frac{f'(x_0)g(x_0) - f(x_0)g'(x_0)}{[g(x_0)]^2}.$$

Earlier in this chapter we considered $f(x) = \sin \frac{1}{x}$ for $x \neq 0$. At that time we assumed some facts concerning the differentiation of the sum, product, and composition of differentiable functions. Now we shall justify our assumptions and consider the differentiability of the composition of differentiable functions. The next theorem is often referred to as the *Chain Rule*.

4.4 THEOREM (CHAIN RULE) Suppose $f: D \to R$ and $g: D' \to R$ with $f(D) \subset D'$. If f is differentiable at x_0 and g is differentiable at $f(x_0)$, then $g \circ f$ is differentiable at x_0 and

$$(g \circ f)'(x_0) = g'(f(x_0))f'(x_0).$$

(Remember that the hypothesis that f is differentiable at x_0 implies that x_0 is an accumulation point of D, and similarly, g differentiable at $f(x_0)$ implies that $f(x_0)$ is an accumulation point of D'.)

Proof. One might be tempted to prove this theorem by writing

$$\frac{(g \circ f)(x) - (g \circ f)(x_0)}{x - x_0} = \frac{g(f(x)) - g(f(x_0))}{f(x) - f(x_0)} \frac{f(x) - f(x_0)}{x - x_0}$$

and claiming that the right-hand side has $g'(f(x_0))f'(x_0)$ as a limit and, hence, $g \circ f$ is differentiable at x_0 as claimed. However, $x \neq x_0$ need not imply that $f(x) \neq f(x_0)$. Hence the denominator may well be zero in the first factor of the right-hand side of the equation. We shall seek to circumvent this trap.

Let $y_0 = f(x_0)$ and, for each $y \in D'$, define

$$h(y) = \frac{g(y) - g(y_0)}{y - y_0} \quad \text{if } y \neq y_0$$

and

$$h(y_0) = g'(y_0).$$

Since g is differentiable at y_0, h has a limit at y_0 and

$$\lim_{y \to y_0} h(y) = g'(y_0) = h(y_0).$$

Thus h is continuous at y_0. Since f is differentiable at x_0 and $y_0 = f(x_0)$, then $h \circ f$ is continuous at x_0. The function T defined by

$$T(x) = \frac{f(x) - f(x_0)}{x - x_0} \quad \text{for } x \in D,\ x \neq x_0$$

has a limit at x_0, namely, $f'(x_0)$. Now for all $x \in D$, $x \neq x_0$,

$$\frac{g(f(x)) - g(f(x_0))}{x - x_0} = (h \circ f)(x) \cdot T(x).$$

You should check the case where $f(x) = f(x_0)$. Since both $h \circ f$ and T have limits at x_0, $g \circ f$ is differentiable at x_0 and

$$\begin{aligned}(g \circ f)'(x_0) &= \lim_{x \to x_0} \frac{g(f(x)) - g(f(x_0))}{x - x_0} = \lim_{x \to x_0} (h \circ f)(x) \cdot T(x) \\ &= (h \circ f)(x_0) \cdot f'(x_0) = g'(f(x_0)) \cdot f'(x_0).\end{aligned}$$

Let us consolidate a few results at this stage of the game. First of all, any constant function is differentiable and has derivative zero everywhere. The function $f(x) = x$ is differentiable everywhere and $f'(x) = 1$ for all x. An easy application of Theorem 4.3 shows that every polynomial function is differentiable and that every rational function (the quotient of two polynomial functions) is differentiable at each point where the denominator is nonzero. Since it is so easy to do and we already know the answer, let us prove now that if $f(x) = x^n$ for n an integer, then $f'(x) = nx^{n-1}$. (Of course, if $n < 0$, we must assume $x \neq 0$, and if $n = 0$, then $f'(x) = 0$ for all x.)

4.5 THEOREM If n is an integer and $f(x) = x^n$ for all x, then f is differentiable for all x if $n > 0$ and for all $x \neq 0$ if $n < 0$, and $f'(x) = nx^{n-1}$. If $n = 0$, then $f'(x) = 0$ for all x.

Proof. For $n = 0$, $f(x) = 1$ for all x; hence, f is differentiable for all x and $f'(x) = 0$, and the theorem is true. We shall now prove the theorem by induction for $n > 0$ and appeal to Theorem 4.3 to extend to the negative integers. The fact that $f(x) = x^n$ is differentiable for each positive integer n follows from Theorem 4.3 by induction, but this is the type of induction proof that should be unnecessary for one who has progressed this far. We shall now proceed to show that $f'(x) = nx^{n-1}$.

For $n = 1$, $f(x) = x$ and $f'(x) = 1 = 1x^0$, so the theorem holds. Suppose the theorem holds for $n = r$; that is, if $f(x) = x^r$, then

$$f'(x) = rx^{r-1}.$$

Consider $g(x) = x^{r+1}$. We may consider this as the product of two functions: $f(x) = x^r$ and $h(x) = x$. By Theorem 4.3,

$$\begin{aligned} g'(x) = f(x)h'(x) + f'(x)h(x) &= x^r(1) + (rx^{r-1})x \\ &= x^r + rx^r = (r+1)x^r. \end{aligned}$$

Thus, the theorem holds for $n = r + 1$, and the induction is complete.

It now remains to verify the conclusion for $f(x) = x^n$ for n a negative integer, $x \neq 0$. Now, if $n < 0$, then $-n > 0$, and $x^n = \dfrac{1}{x^{-n}}$ is differentiable for $x \neq 0$ by our remarks earlier in this proof and Theorem 4.3. Moreover, if $f(x) = x^n = \dfrac{1}{x^{-n}}$, then

$$f'(x) = \frac{-(-n)x^{-n-1}}{(x^{-n})^2} = nx^{n-1}.$$

This concludes the proof.

■ **Example 4.5** Let us examine the use of Theorems 4.3, 4.4, and 4.5 in finding the derivative of a function such as $F(x) = \sqrt{1 + 3x^2}$, which is the composition of the function $f(x) = 1 + 3x^2$ and the function $g(u) = \sqrt{u}$. For each x, f is differentiable at x and, using Theorems 4.3 and 4.4, $f'(x) = 6x$. Moreover, for each x, $f(x) > 0$; hence g is differentiable at $f(x)$. Note that $g'(u) = \dfrac{1}{2\sqrt{u}}$. Thus, by the Chain Rule, $g \circ f = F$ is differentiable at x and

$$F'(x) = g'(f(x)) \cdot f'(x) = \frac{1}{2\sqrt{1 + 3x^2}} \cdot 6x. \qquad ■$$

4.3 ROLLE'S THEOREM AND THE MEAN-VALUE THEOREM

The classical maxima and minima problems from the calculus are no doubt familiar to the reader. The fact to be applied here is roughly that the graph of a differentiable function has a horizontal tangent at a maximum or minimum point. This is not quite accurate, but let us define some terms before proceeding further.

DEFINITION Let $f: D \to R$. A point $x_0 \in D$ is a *relative maximum* (*minimum*) of f iff there is a neighborhood Q of x_0 such that if $x \in Q \cap D$, then

$$f(x) \leqslant f(x_0) \quad (f(x) \geqslant f(x_0)).$$

Consider a few examples and reconsider the statement preceding this definition.

■ **Example 4.6** Perhaps the point is most easily made by looking at the function $f: [0, 1] \to R$ such that $f(x) = x$. Since the domain of f is the interval $[0, 1]$, f has a minimum at 0 and a maximum at 1; indeed, 0 is a relative minimum of f, and 1 is a relative maximum of f. Since $f'(x) = 1$ for all $x \in [0, 1]$, it is clear what is wrong with our statement concerning the existence of a horizontal tangent at a maximum or minimum point. We shall correct this defect in the following theorem. ■

4.6 THEOREM Suppose $f: [a, b] \to R$ and suppose f has either a relative maximum or a relative minimum at $x_0 \in (a, b)$. If f is differentiable at x_0, then $f'(x_0) = 0$.

Proof. Assume f has a relative maximum at x_0. Then there is $\delta > 0$ such that, if $x_0 - \delta < x < x_0 + \delta$, then $x \in [a, b]$ [since $x_0 \in (a, b)$] and $f(x) \leqslant f(x_0)$. Consider any sequence $\{x_n\}_{n=1}^{\infty}$ converging to x_0 such that $x_0 - \delta < x_n < x_0$. Then, since f is differentiable at x_0,

$$\left\{\frac{f(x_n) - f(x_0)}{x_n - x_0}\right\}_{n=1}^{\infty}$$

converges to $f'(x_0)$. But

$$\frac{f(x_n) - f(x_0)}{x_n - x_0} \geqslant 0$$

for each n since $f(x_n) \leqslant f(x_0)$ and $x_n < x_0$, and hence $f'(x_0) \geqslant 0$. Consider now a sequence $\{y_n\}_{n=1}^{\infty}$ such that $x_0 < y_n < x_0 + \delta$. As before,

$$\left\{\frac{f(y_n) - f(x_0)}{y_n - x_0}\right\}_{n=1}^{\infty}$$

converges to $f'(x_0)$, but this time

$$\frac{f(y_n) - f(x_0)}{y_n - x_0} \leqslant 0;$$

hence $f'(x_0) \leq 0$. Therefore, $f'(x_0) = 0$. The case in which f has a relative minimum at x_0 is left to the reader.

The reader should now realize that an important fact in this proof is that we are free to choose points in the domain on either side of x_0 as close as we please. Since both the maximum and the minimum occurred at end points in the preceding example, the theorem did not apply.

The following theorem, known as Rolle's Theorem, is an application of Theorem 4.6. We shall postpone a discussion of its geometric interpretation until we have completed the proof.

4.7 ROLLE'S THEOREM Suppose $f: [a, b] \to R$ is continuous on $[a, b]$ and f is differentiable on (a, b). Then if $f(a) = f(b) = 0$, there is $c \in (a, b)$ such that $f'(c) = 0$.

Proof. If $f(x) = 0$ for all $x \in [a, b]$, then $f'(x) = 0$ for all $x \in [a, b]$, and the theorem is proved.

Suppose $f(x) \neq 0$ for some $x \in [a, b]$. By Theorem 3.7, $[a, b]$ is compact since it is closed and bounded; hence, by Corollary 3.11, f assumes its maximum and minimum in $[a, b]$, say at x_1 and x_2, respectively. Since f is not identically zero on $[a, b]$ and $f(a) = f(b) = 0$, at least one of x_1 and x_2 must belong to (a, b), say $x_1 \in (a, b)$. Now by Theorem 4.6, $f'(x_1) = 0$.

In essence, the theorem states that if the graph of a differentiable function touches the x-axis at a and at b, then somewhere between a and b there is a horizontal tangent. See Figure 4.1.

Note that the theorem states that there is at least one c such that $f'(c) = 0$. There may be more such points as indicated in Figure 4.1.

If we examine the graph in Figure 4.1 from a geometric point of view—that is, ignoring the choice of coordinate system—it would appear that the tangent line at c is parallel to the line connecting the endpoints of the curve. This is essentially the content of the Mean-Value Theorem to follow.

Suppose we have a smooth curve, the graph of f, connecting the points $(a, f(a))$ and $(b, f(b))$. We can rotate and translate the coordinate axes so that both the points lie on the new x-axis. Then Rolle's Theorem will guarantee a point on the curve where the tangent line is parallel to the new x-axis, which

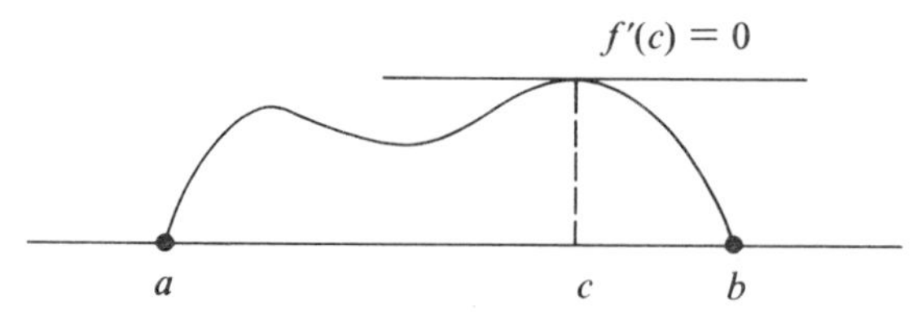

Figure 4.1

contains the segment connecting the points $(a, f(a))$ and $(b, f(b))$. See Figure 4.2.

The slope of the segment joining $(a, f(a))$ and $(b, f(b))$ is

$$\frac{f(b) - f(a)}{b - a},$$

so this is the slope of the tangent line at c; that is,

$$f'(c) = \frac{f(b) - f(a)}{b - a}.$$

The precise statement and proof follow.

4.8 MEAN-VALUE THEOREM If f: $[a, b] \to R$ is continuous on $[a, b]$ and differentiable on (a, b), then there is a $c \in (a, b)$ such that

$$f'(c) = \frac{f(b) - f(a)}{b - a}.$$

Proof. To prove this theorem, we shall find a linear function L such that $f - L$ satisfies Rolle's Theorem and then apply Rolle's Theorem to obtain the desired result. This corresponds to the change of coordinate system mentioned before this theorem. The function L must be linear and satisfy

$$L(a) = f(a) \quad \text{and} \quad L(b) = f(b).$$

This is accomplished by choosing

$$L(x) = \left[\frac{f(b) - f(a)}{b - a}\right](x - a) + f(a),$$

the equation of the line passing through $(a, f(a))$ and $(b, f(b))$. L is continuous and differentiable everywhere and

$$L'(x) = \frac{f(b) - f(a)}{b - a}.$$

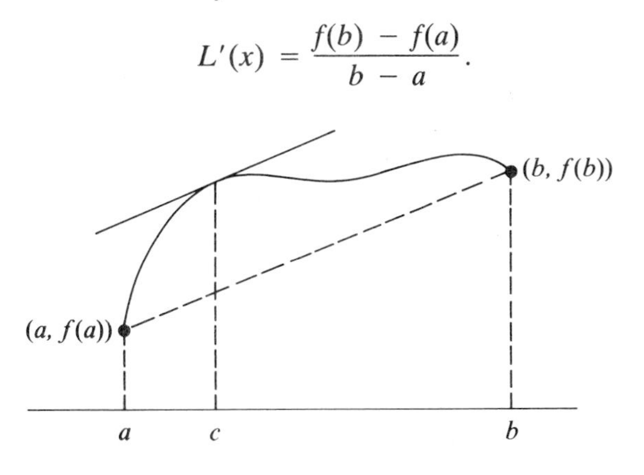

Figure 4.2

Let $g = f - L$. The function g satisfies the hypotheses of Rolle's Theorem, hence there is $c \in (a, b)$ such that

$$0 = g'(c) = f'(c) - L'(c) = f'(c) - \frac{f(b) - f(a)}{b - a}.$$

Thus

$$f'(c) = \frac{f(b) - f(a)}{b - a}.$$

The Mean-Value Theorem has some very important applications in relating the behavior of f and f'. The next theorem will serve to illustrate this.

4.9 THEOREM Suppose f is continuous on $[a, b]$ and differentiable on (a, b). Under these hypotheses:

i. If $f'(x) \neq 0$ for all $x \in (a, b)$, then f is 1–1.
ii. If $f'(x) = 0$ for all $x \in (a, b)$, then f is constant.
iii. If $f'(x) > 0$ for all $x \in (a, b)$, then $x < y$ and $x, y \in [a, b]$ imply $f(x) < f(y)$ (that is, f is strictly increasing).
iv. If $f'(x) < 0$ for all $x \in (a, b)$, then $x < y$ and $x, y \in [a, b]$ imply $f(x) > f(y)$ (f is strictly decreasing).

Proof. Consider any $x, y \in [a, b]$ with $x < y$. Now f is continuous on $[x, y]$ and differentiable on (x, y); so, by the Mean-Value Theorem, there is $c \in (x, y)$ such that

$$\frac{f(x) - f(y)}{x - y} = f'(c);$$

that is, $f(x) - f(y) = f'(c)(x - y)$. With this fact in mind, we shall proceed.

i. Suppose f is not 1–1. Then there are $x, y \in [a, b]$ with $x < y$ such that $f(x) = f(y)$. Thus, there is $c \in (x, y)$ such that

$$f'(c) = \frac{f(x) - f(y)}{x - y} = 0,$$

contrary to $f'(c) \neq 0$ for all $c \in [a, b]$.

ii. Suppose f is not constant on $[a, b]$. Then there are $x, y \in [a, b]$, $x < y$, such that $f(x) \neq f(y)$. There is $c \in (a, b)$ such that

$$f'(c) = \frac{f(x) - f(y)}{x - y} \neq 0$$

contrary to $f'(c) = 0$ for all $c \in [a, b]$.

iii. Suppose $x < y$ and $x, y \in [a, b]$. There is $c \in (x, y)$ such that

$$\frac{f(x) - f(y)}{x - y} = f'(c) > 0;$$

hence $f(x) < f(y)$.

iv. Suppose $x < y$ and $x, y \in [a, b]$. There is $c \in (x, y)$ such that

$$\frac{f(x) - f(y)}{x - y} = f'(c) < 0;$$

hence $f(x) > f(y)$.

Now that the proof has been presented, it is clear that parts (iii) and (iv) can be changed as follows to give additional information:

iii. $f'(x) \geqslant 0$ for all $x \in (a, b)$ implies f is increasing.
iv. $f'(x) \leqslant 0$ for all $x \in (a, b)$ implies f is decreasing.

As an obvious corollary to Theorem 4.9, we have the following theorem.

4.10 THEOREM Suppose that f and g are continuous on $[a, b]$ and differentiable on (a, b) and that $f'(x) = g'(x)$ for all $x \in (a, b)$. Then there is a real number k such that

$$f(x) = g(x) + k$$

for all $x \in [a, b]$.

Proof. Consider $h(x) = f(x) - g(x)$ for all $x \in [a, b]$. Clearly, h is continuous on $[a, b]$ and differentiable on (a, b) and

$$h'(x) = f'(x) - g'(x) = 0$$

for all $x \in (a, b)$. Hence, by Theorem 4.9, there is a real number k such that

$$k = h(x) = f(x) - g(x)$$

for all $x \in [a, b]$. Thus, $f(x) = g(x) + k$ for all $x \in [a, b]$.

■ **Example 4.7** The Mean-Value Theorem has a variety of uses, one of which is that of estimating values of certain functions. Suppose $p > 1$ and $h > 0$. Define the function $f(x) = (1 + x)^p$. Assuming the basic facts about the differentiation of powers, f is differentiable for $x > 0$. So, if $h > 0$, there is t such that $0 < t < h$ and

$$f'(t) = \frac{f(h) - f(0)}{h - 0}$$

or

$$f(h) = f'(t)(h - 0) + f(0) = p(1 + t)^{p-1}h + 1.$$

Now $p > 1$, hence $(1 + t)^{p-1} > 1$, so

$$f(h) = (1 + h)^p = p(1 + t)^{p-1}h + 1 > ph + 1.$$

See Exercise 19 for the case of $p < 1$. ■

Consider a function f differentiable on $[a, b]$. If $f'(x) \neq 0$ for all $x \in (a, b)$, then by Theorem 4.9, f is 1–1 and by Theorem 3.16, f is monotone. If f is increasing, then, for all $x, y \in [a, b]$ with $x \neq y$, we have

$$\frac{f(x) - f(y)}{x - y} > 0.$$

Hence $f'(x) \geqslant 0$ for all $x \in [a, b]$. Similarly, if f is decreasing, $f'(x) \leqslant 0$ for all $x \in [a, b]$. We have thus proved the following lemma.

LEMMA If f is differentiable on $[a, b]$ and $f'(x) \neq 0$ for all $x \in (a, b)$, then either $f'(x) \geqslant 0$ for all $x \in [a, b]$ or $f'(x) \leqslant 0$ for all $x \in [a, b]$.

Let us consider this result from a slightly different point of view. Suppose f is differentiable on $[a, b]$ and there are $x, y \in [a, b]$ such that $f'(x) > 0$ and $f'(y) < 0$. If we restrict our attention to $[x, y]$ (assuming $x < y$), then, by the lemma, there is $c \in (x, y)$ such that $f'(c) = 0$. Thus, we are led to suspect that f' has the intermediate-value property. That this is indeed the case is stated in the following theorem.

4.11 THEOREM Suppose f is differentiable on $[a, b]$ and λ is a real number such that $f'(a) < \lambda < f'(b)$ or $f'(b) < \lambda < f'(a)$. Then there is $c \in (a, b)$ such that $f'(c) = \lambda$.

Proof. Define

$$g(x) = f(x) - \lambda x$$

for all $x \in [a, b]$. Then g is differentiable on $[a, b]$ and $g'(x) = f'(x) - \lambda$. If $f'(a) < \lambda < f'(b)$, then $g'(a) < 0$ and $g'(b) > 0$; if $f'(b) < \lambda < f'(a)$, then $g'(a) > 0$ and $g'(b) < 0$. In either case, g' has opposite signs at a and at b. Hence, by the lemma, there is $c \in (a, b)$ such that

$$0 = g'(c) = f'(c) - \lambda$$

so that $f'(c) = \lambda$.

Suppose f: $[a, b] \to R$ is such that there is a differentiable function g: $[a, b] \to R$ such that $g'(x) = f(x)$ for all $x \in [a, b]$. Now f need not be a continuous fashion, but f must have the intermediate-value property; in other words, if $x, y \in [a, b]$ and $f(x) < \lambda < f(y)$, then there is c between x and y such that $f(c) = \lambda$. This means that a function that is a derivative of some function must be rather special. In particular, $f(x) = [x]$ cannot be the derivative of a function.

The following theorem is a generalization of the Mean-Value Theorem.

4.12 CAUCHY MEAN-VALUE THEOREM If f and g are continuous on $[a, b]$ and differentiable on (a, b), then there is $c \in (a, b)$ such that

$$[f(b) - f(a)]g'(c) = [g(b) - g(a)]f'(c).$$

Proof. Define $h(t) = [f(b) - f(a)]g(t) - [g(b) - g(a)]f(t)$ for each $t \in [a, b]$. Note that h is continuous on $[a, b]$ and differentiable on (a, b) and that $h(a) = h(b)$. So, by the Mean-Value Theorem, there is $c \in (a, b)$ such that

$$h'(c) = [f(b) - f(a)]g'(c) - [g(b) - g(a)]f'(c) = 0.$$

4.4 L'HOSPITAL'S RULE AND THE INVERSE-FUNCTION THEOREM

We can use Cauchy's Mean-Value Theorem to prove a familiar rule from calculus, L'Hospital's Rule.

4.13 L'HOSPITAL'S RULE Suppose f and g are continuous on $[a, b]$ and differentiable on (a, b). If $x_0 \in [a, b]$,

1. $g'(x) \neq 0$ for all $x \in [a, b]$, $x \neq x_0$,
2. $f(x_0) = g(x_0) = 0$, and
3. $\dfrac{f'}{g'}$ has a limit at x_0,

then $\dfrac{f}{g}$ has a limit at x_0 and

$$\lim_{x \to x_0} \frac{f}{g}(x) = \lim_{x \to x_0} \frac{f'}{g'}(x).$$

Proof. Let $\{x_n\}_{n=1}^{\infty}$ be any sequence converging to x_0 with $x_n \in [a, b] \setminus \{x_0\}$ for all n. By the Cauchy Mean Value Theorem, there is a sequence $\{c_n\}_{n=1}^{\infty}$ such that

i. c_n is between x_n and x_0 for each n, and

ii. $[f(x_n) - f(x_0)]g'(c_n) = [g(x_n) - g(x_0)]f'(c_n)$ for each n.

Since $x_n \neq x_0$ and $g'(x) \neq 0$ for $x \neq x_0$, Rolle's Theorem tells us that $g(x_n) \neq g(x_0) = 0$ for all n. Since $f(x_0) = g(x_0) = 0$,

$$\frac{f(x_n)}{g(x_n)} = \frac{f(x_n) - f(x_0)}{g(x_n) - g(x_0)} = \frac{f'(c_n)}{g'(c_n)},$$

and since $\dfrac{f'}{g'}$ has a limit at x_0, the sequence $\left\{\dfrac{f'(c_n)}{g'(c_n)}\right\}_{n=1}^{\infty}$ converges; hence, so does the sequence $\left\{\dfrac{f(x_n)}{g(x_n)}\right\}_{n=1}^{\infty}$ and both to the same limit.

Therefore $\dfrac{f}{g}$ has a limit at x_0 and

$$\lim_{x \to x_0} \left(\frac{f}{g}\right)(x) = \lim_{x \to x_0} \left(\frac{f'}{g'}\right)(x).$$

■ **Example 4.8** To illustrate the use of L'Hospital's Rule, consider the functions f and g defined as follows:

$$f(x) = \sqrt{x} - \sqrt{2} + \sqrt{x - 2}$$

and

$$g(x) = \sqrt{x^2 - 4}, \quad \text{for all } x \in [2, 3].$$

Now f and g are both continuous on $[2, 3]$ and differentiable on $(2, 3)$, and $f(2) = g(2) = 0$. Further

$$f'(x) = \frac{1}{2\sqrt{x}} + \frac{1}{2\sqrt{x - 2}} \quad \text{and} \quad g'(x) = \frac{x}{\sqrt{x^2 - 4}}.$$

So

$$\begin{aligned} \frac{f'(x)}{g'(x)} &= \frac{(\sqrt{x - 2} + \sqrt{x})\sqrt{x - 2}\sqrt{x + 2}}{(2\sqrt{x}\sqrt{x - 2})(x)} \\ &= \frac{(\sqrt{x - 2} + \sqrt{x})(\sqrt{x + 2})}{2x\sqrt{x}}. \end{aligned}$$

It is clear that $\dfrac{f'}{g'}$ has a limit at 2 and that limit is $\dfrac{1}{2}$. So, by L'Hospital's Rule, $\dfrac{f}{g}$ has a limit at 2 and the limit is $\dfrac{1}{2}$. ■

The next theorem is a one-dimensional version of the Inverse-Function Theorem. In Chapter 3, we observed that, if f: $[a, b] \to R$ is continuous and 1–1, then f^{-1} is also continuous. It is now natural to inquire what can be obtained if it is assumed that f is differentiable. Before proceeding further, let's look at the possibilities. If f is 1–1 and differentiable, and if f^{-1} is differentiable, then, by the Chain Rule,

$$(f^{-1} \circ f)'(x) = (f^{-1})'(f(x))f'(x),$$

but $(f^{-1} \circ f)(x) = x$ for all $x \in [a, b]$; hence

$$1 = (f^{-1} \circ f)'(x) = (f^{-1})'(f(x))f'(x)$$

for all $x \in [a, b]$. This last equation shows that, for $x \in [a, b]$,

$$(f^{-1})'(f(x)) = \frac{1}{f'(x)}$$

and, in particular, $f'(x) \neq 0$. These observations give us some insight into what hypotheses are needed and what conclusions might be drawn in the setting mentioned above.

4.14 THEOREM Suppose $f: [a, b] \to R$ is continuous and differentiable with $f'(x) \neq 0$ for all $x \in [a, b]$. Then f is 1–1, f^{-1} is continuous and differentiable on $f([a, b])$, and

$$(f^{-1})'(f(x)) = \frac{1}{f'(x)}$$

for all $x \in [a, b]$.

Proof. Since $f'(x) \neq 0$ for all $x \in [a, b]$, then, by Theorem 4.9, f is 1–1. Let us suppose that $f([a, b]) = [c, d]$. (The fact that f is continuous guarantees that $f([a, b])$ is indeed a closed interval.) We shall now proceed to show that $f^{-1}: [c, d] \to [a, b]$ is differentiable. Choose $y_0 \in [c, d]$ and $\{y_n\}_{n=1}^{\infty}$ any sequence in $[c, d] \setminus \{y_0\}$ converging to y_0. Let

$$x_n = f^{-1}(y_n)$$

for $n = 0, 1, 2, \ldots$. By Theorem 3.12, f^{-1} is continuous and $\{x_n\}_{n=1}^{\infty}$ converges to $x_0 = f^{-1}(y_0)$; and, since f^{-1} is 1–1, $x_n \neq x_0$ for all n. By the differentiability of f,

$$\left\{\frac{f(x_n) - f(x_0)}{x_n - x_0}\right\}_{n=1}^{\infty}$$

converges to $f'(x_0)$. By hypothesis, $f'(x_0) \neq 0$ and, since f is 1–1,

$$\frac{f(x_n) - f(x_0)}{x_n - x_0} \neq 0$$

for $n = 1, 2, \ldots$. Hence

$$\left\{\frac{f^{-1}(y_n) - f^{-1}(y_0)}{y_n - y_0}\right\}_{n=1}^{\infty} = \left\{\frac{x_n - x_0}{f(x_n) - f(x_0)}\right\}_{n=1}^{\infty}$$

converges to $\dfrac{1}{f'(x_0)}$. Thus, by Theorem 4.1, f^{-1} is differentiable and

$$(f^{-1})'(f(x_0)) = \frac{1}{f'(x_0)}.$$

See Exercise 29 for a variation on this theorem.

Example 4.9 We close this chapter with a few applications of our results. Consider the function $g(x) = x^2$ defined for all real numbers x. Of course, g is differentiable and $g'(x) = 2x$. We shall restrict our attention to the behavior of g on the interval $[8, 10]$. If $x \in [8, 10]$, then $g'(x) = 2x \neq 0$, and, by Theorem 4.14, g is 1–1 on $[8, 10]$, g^{-1} is differentiable on $[64, 100]$, and $(g^{-1})'(g(x)) = \frac{1}{g'(x)}$. It is customary to write the inverse of g as $\sqrt{x}$ where, of course, $\sqrt{x}$ denotes the positive square root of x. Now, $g([8, 10]) = [64, 100]$, and, if $h(x) = \sqrt{x} = g^{-1}(x)$, we have

$$h: [64, 100] \to [8, 10].$$

Also,

$$h'(x) = h'(g(h(x))) = \frac{1}{g'(h(x))} = \frac{1}{2\sqrt{x}}$$

for each $x \in [64, 100]$, by Theorem 4.14. Suppose one wants to approximate $\sqrt{80} = h(80)$. Since $h(81) = 9$ is easy to calculate, we shall try to use this result to come close to $\sqrt{80}$. By the Mean-Value Theorem,

$$h(81) - h(80) = h'(x_0)(81 - 80)$$

for some $x_0 \in [80, 81]$. Thus

$$h(80) = h(81) - h'(x_0) = 9 - \frac{1}{2\sqrt{x_0}}.$$

Although we do not know the exact value of x_0, we do know that $8 < \sqrt{x_0} < 9$. We may conclude that

$$9 - \frac{1}{18} < \sqrt{80} < 9 - \frac{1}{16}.$$ ■

Example 4.10 We next consider the function $f(x) = \tan x$ with domain $\left(-\frac{\pi}{2}, \frac{\pi}{2}\right)$. Assume that f is differentiable, $f'(x) = \sec^2 x$, and $\sec^2 x \neq 0$ for all $x \in \left(-\frac{\pi}{2}, \frac{\pi}{2}\right)$. Further assume the identity $\sec^2 x = 1 + \tan^2 x$. Since $f'(x) \neq 0$ for all $x \in \left(-\frac{\pi}{2}, \frac{\pi}{2}\right)$, then f is 1–1 and has an inverse g, usually written $g(x) = \text{Arctan } x$. By the Inverse-Function Theorem, g is differentiable and

$$\begin{aligned} g'(x) = g'(f(g(x))) &= \frac{1}{f'(g(x))} = \frac{1}{\sec^2(g(x))} = \frac{1}{1 + \tan^2(g(x))} \\ &= \frac{1}{1 + (f(g(x)))^2} = \frac{1}{1 + x^2}. \end{aligned}$$

Of course, this is the usual formula for the derivative of Arctan x. ■

EXERCISES

4.1 THE DERIVATIVE OF A FUNCTION

1. Use the definition to find the derivative of $f(x) = \sqrt{x}$, for $x > 0$. If f differentiable at 0? Explain.
2. Use the definition to find the derivative of $g(x) = x^2$.
3. Define $h(x) = x^3 \sin \frac{1}{x}$ for $x \neq 0$ and $h(0) = 0$. Show that h is differentiable everywhere and that h' is continuous everywhere but fails to have a derivative at one point. You may use the rules for differentiating products, sums, and quotients of elementary functions that you learned in calculus.
4. Suppose $f\colon (a, b) \to R$ is differentiable at $x \in (a, b)$. Prove that

$$\lim_{h \to 0} \frac{f(x + h) - f(x - h)}{2h}$$

exists and equals $f'(x)$. Give an example of a function where this limit exists but the function is not differentiable.
5. A function $f\colon (a, b) \to R$ satisfies a *Lipschitz condition* at $x \in (a, b)$ iff there is $M > 0$ and $\epsilon > 0$ such that $|x - y| < \epsilon$ and $y \in (a, b)$ imply that $|f(x) - f(y)| \leq M|x - y|$. Give an example of a function that fails to satisfy a Lipschitz condition at a point of continuity. If f is differentiable at x, prove that f satisfies a Lipschitz condition at x.
6. A function $f\colon (a, b) \to R$ is said to be uniformly differentiable iff f is differentiable on (a, b) and for each $\epsilon > 0$, there is $\delta > 0$ such that $0 < |x - y| < \delta$ and $x, y \in (a, b)$ imply that

$$\left| \frac{f(x) - f(y)}{x - y} - f'(x) \right| < \epsilon.$$

Prove that if f is uniformly differentiable on (a, b), then f' is continuous on (a, b).

*7. Suppose $f\colon (a, b) \to R$ is continuous on (a, b) and differentiable at $x_0 \in (a, b)$. Define

$$g(x) = \frac{f(x) - f(x_0)}{x - x_0} \quad \text{for } x \in (a, b) \setminus \{x_0\},$$
$$g(x_0) = f'(x_0).$$

Prove that g is continuous on (a, b).

8. Suppose that f, g, and h have domain D, $x_0 \in D$, and x_0 is an accumulation point of D. Assume that f and h are differentiable at x_0, $f'(x_0) = h'(x_0)$, $\lim_{x \to x_0} f(x) = \lim_{x \to x_0} h(x)$, and $f(x) \leq g(x) \leq h(x)$ for all x in D. Prove that g is differentiable at x_0 and that $g'(x_0) = f'(x_0) = h'(x_0)$. (This problem was suggested by Clyde Dubbs of New Mexico Institute of Mining and Technology.)

4.2 THE ALGEBRA OF DERIVATIVES

9. Prove $f\colon (0, 1) \to R$ defined by $f(x) = \sqrt{2x^2 - 3x + 6}$ is differentiable on $(0, 1)$ and compute the derivative.
10. Suppose $f\colon [a, b] \to [c, d]$, $g\colon [c, d] \to [p, q]$, and $h\colon [p, q] \to R$, with f differentiable at $x_0 \in [a, b]$, g differentiable at $f(x_0)$, and h differentiable at $g(f(x_0))$. Prove that $h \circ (g \circ f)$ is differentiable at x_0 and find the derivative.

11. Suppose f: $[a, b] \to [c, d]$ and g: $[c, d] \to R$ are differentiable on $[a, b]$ and $[c, d]$, respectively. Suppose

$$f': [a, b] \to R \quad \text{and} \quad g': [c, d] \to R$$

are also differentiable on $[a, b]$ and $[c, d]$, respectively. Show that $(g \circ f)'$: $[a, b] \to R$ is differentiable and find the derivative.

12. Suppose f: $R \to R$ is differentiable and define $g(x) = x^2 f(x^3)$. Show that g is differentiable and compute g'.

13. Define $f(x) = \sqrt{x + \sqrt{x + \sqrt{x}}}$ for $x \geqslant 0$. Determine where f is differentiable and compute the derivative.

4.3 ROLLE'S THEOREM AND THE MEAN-VALUE THEOREM

14. Define f: $[0, 2] \to R$ by $f(x) = \sqrt{2x - x^2}$. Show that f satisfies the conditions of Rolle's Theorem and find c such that $f'(c) = 0$.

15. Define f: $R \to R$ by $f(x) = \dfrac{1}{1 + x^2}$. Prove that f has a maximum value and find the point at which that maximum occurs.

16. Prove that the equation $x^3 - 3x + b = 0$ has at most one root in the interval $[-1, 1]$.

17. Use the Mean-Value Theorem to prove that

$$ny^{n-1}(x - y) \leqslant x^n - y^n \leqslant nx^{n-1}(x - y)$$

if $n \geqslant 1$ and $0 \leqslant y \leqslant x$.

18. Use the Mean-Value Theorem to prove that

$$\sqrt{1 + h} < 1 + \frac{1}{2}h \quad \text{for } h > 0.$$

19. Generalize Exercise 18 as follows: If $0 < p < 1$ and $h > 0$, then show that

$$(1 + h)^p < 1 + ph.$$

You may assume the usual rules about differentiating powers.

20. Suppose f: $(a, b) \to R$ is differentiable and $|f'(x)| \leqslant M$ for all $x \in (a, b)$. Prove that f is uniformly continuous on (a, b). Give an example of a function f: $(0, 1) \to R$ that is differentiable and uniformly continuous on $(0, 1)$ but such that f' is unbounded.

21. Suppose f is differentiable on (a, b), except possibly at $x_0 \in (a, b)$, and continuous on $[a, b]$; assume $\lim_{x \to x_0} f'(x)$ exists. Prove that f is differentiable at x_0 and f' is continuous at x_0.

22. Define $f(x) = x + 2x^2 \sin \dfrac{1}{x}$ for $x \neq 0$ and $f(0) = 0$. Prove that f is differentiable everywhere. Show that there exists a number a such that $f'(a) > 0$ but there does not exist a neighborhood of a in which f is increasing.

23. Prove that the function $f(x) = 2x^3 + 3x^2 - 36x + 5$ is 1–1 on the interval $[-1, 1]$. Is f increasing or decreasing?

24. Show that the function $f(x) = x^3 - 3x^2 + 17$ is not 1–1 on the interval $[-1, 1]$.

25. Give an example of a function $f: R \to R$ that is differentiable and 1–1, but $f'(x) = 0$ for some $x \in R$.

26. If $f: [a, b] \to R$ is differentiable at c, $a < c < b$, and $f'(c) > 0$, prove that there is x, $c < x < b$, such that $f(x) > f(c)$.

4.4 L'HOSPITAL'S RULE AND THE INVERSE-FUNCTION THEOREM

27. Assume the rules for differentiating the elementary functions and use L'Hospital's Rule to find the following limits:

a. $\lim_{x \to 1} \dfrac{\ln x}{x - 1}$ b. $\lim_{x \to 0} \dfrac{x}{e^x - 1}$ c. $\lim_{x \to 0} \dfrac{\sin x}{x}$

28. Use L'Hospital's Rule to find the limit:

$$\lim_{x \to 0} \frac{x^2 \sin x}{\sin x - x \cos x}.$$

You will want to use the rule several times, but be sure to justify each use.

***29.** Prove the following variant of Theorem 4.14. Suppose $f: [a, b] \to R$ is 1–1. If f is differentiable at $c \in [a, b]$, $f'(c) \neq 0$, and f^{-1} is continuous at $d = f(c)$, then f^{-1} is differentiable at d and

$$(f^{-1})'(d) = \frac{1}{f'(c)}.$$

30. Find an equation for the tangent line to the graph of f^{-1} at the point (3, 1) if $f(x) = x^3 + 2x^2 - x + 1$.

31. Use the Inverse-Function Theorem to derive the formula for the derivative of the inverse of $\sin x$ on the interval $\left[-\frac{\pi}{2}, \frac{\pi}{2}\right]$. You may assume the usual facts about the function $f(x) = \sin x$.

32. Suppose both f and f^{-1} are twice-differentiable functions. Derive a formula for $(f^{-1})''$.

MISCELLANEOUS

33. Suppose $f: (a, b) \to R$ is differentiable at $x_0 \in (a, b)$ with $\{\alpha_n\}_{n=1}^{\infty}$ and $\{\beta_n\}_{n=1}^{\infty}$ two sequences in $(a, b) \setminus \{x_0\}$ converging to x_0 such that the sequence

$$\left\{\frac{\beta_n - x_0}{\beta_n - \alpha_n}\right\}_{n=1}^{\infty}$$

is bounded. Prove that

$$\left\{\frac{f(\beta_n) - f(\alpha_n)}{\beta_n - \alpha_n}\right\}_{n=1}^{\infty}$$

converges to $f'(x_0)$.

34. Suppose $f: R \to R$ is such that $f(x + y) = f(x)f(y)$, f is differentiable at zero, and f is not identically zero. Prove that f is differentiable everywhere and that $f'(x) = f(x)f'(0)$. Assuming the properties of the exponential function, prove that $f(x) = e^{cx}$ where $c = f'(0)$.

35. Suppose $f\colon [a, b] \to R$ is differentiable and f'' exists at $t \in (a, b)$. Prove that

$$f''(t) = \lim_{h \to 0} \frac{f(t+h) - 2f(t) + f(t-h)}{h^2}.$$

Give an example where this limit exists, but f' is not differentiable at t.

36. Suppose $f\colon D \to R$, $g\colon E \to R$, $x_0 \in D \cap E$, x_0 an accumulation point of $D \cap E$. Suppose further that there is $\epsilon > 0$ such that

$$D \cap [x_0 - \epsilon, x_0 + \epsilon] = E \cap [x_0 - \epsilon, x_0 + \epsilon]$$

with $f(x) = g(x)$ for all $x \in D \cap E \cap [x_0 - \epsilon, x_0 + \epsilon]$. Prove that f is differentiable at x_0 iff g is differentiable at x_0.

PROJECT

The purpose of this project is to examine the Newton–Raphson method for solving an equation of the form $f(x) = 0$, establish convergence of the iterative process under certain conditions, and give estimates of the speed of convergence.

Suppose $f\colon [a, b] \to R$ is twice differentiable; that is, f and f' are both differentiable on $[a, b]$. Suppose further that there are m and M, both positive, such that $|f'(x)| \geqslant m$ and $0 < |f''(x)| \leqslant M$ for all $x \in [a, b]$. Finally suppose that $f(a)f(b) < 0$. Define a sequence as follows:

$$a_1 = a \quad \text{if } f'f'' < 0 \qquad \text{and} \qquad a_1 = b \quad \text{if } f'f'' > 0,$$

$$a_n = a_{n-1} - \frac{f(a_{n-1})}{f'(a_{n-1})} \quad \text{for } n \geqslant 2.$$

Then $\{a_n\}_{n=1}^{\infty}$ converges to the unique solution r in $[a, b]$ of the equation $f(r) = 0$. Moreover,

$$|r - a_{n+1}| \leqslant \frac{M(r - a_n)^2}{2m}$$

Look at the graph in Figure 4.3. The basic idea is this: In this case, pick b as the first approximation to r. Then the next approximation is found by taking the point where the tangent line through $(b, f(b))$ intersects the x-axis. A little algebra shows that

$$a_2 = b - \frac{f(b)}{f'(b)}$$

for the function shown in Figure 4.3. In the case indicated, it would appear that the sequence $\{a_n\}_{n=1}^{\infty}$ is decreasing and bounded from below, hence convergent. These impressions will determine our strategy in establishing the truth of the theorem. A preliminary result, a special case of Taylor's Theorem, is necessary. See Section 6.6 for a more general discussion of Taylor's Theorem.

1. Suppose $f\colon [a, b] \to R$ is twice differentiable. Prove that for each x_0, $x \in [a, b]$, there is c between x_0 and x such that

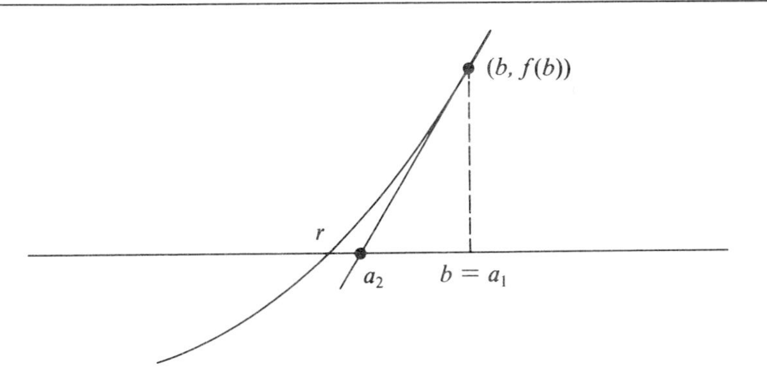

Figure 4.3

$$f(x) = f(x_0) + f'(x_0)(x - x_0) + \frac{f''(c)}{2}(x - x_0)^2.$$

(*Hint:* Define F: $[a, b] \to R$ by

$$F(t) = f(t) + f'(t)(x - t) + M(x - t)^2$$

where M is chosen so that $F(x_0) = f(x)$. Show that $F(x) = F(x_0)$ and apply the Mean-Value Theorem to obtain M.)

Notice that the conditions on f' and f'' ensure that neither changes sign in $[a, b]$, by Theorem 4.11. Also, f is continuous and 1–1 with $f(a)$ and $f(b)$ having opposite sign, hence there is exactly one solution in $[a, b]$ to the equation $f(x) = 0$; we call it r. In what is to follow, we assume $f' > 0$ and $f'' > 0$. The other cases are similar.

2. Show that $r < a_{n+1} < a_n \leq b$ for all $n \in J$. Prove this by induction, using the Taylor formula from part 1. In Taylor's formula, use $r = x$ and $a_n = x_0$. Remember, $f(r) = 0$ and $f''(c) > 0$ for all $c \in [a, b]$.
3. By part 2, $\{a_n\}_{n=1}^{\infty}$ is decreasing and bounded from below by r, hence it converges; call the limit s. Prove that $s = r$. Remember f and f' are both continuous on $[a, b]$ and $f' > 0$ on $[a, b]$.
4. The recurrence relation for $\{a_n\}_{n=1}^{\infty}$ shows that $f(a_n) = f'(a_n)(a_n - a_{n+1})$. Substitute this in Taylor's formula to obtain the inequality

$$|r - a_{n+1}| \leq \frac{M(r - a_n)^2}{2m}.$$

CHAPTER 5

The Riemann Integral

The concept in this chapter is one the reader should have some familiarity with—that of the Riemann integral. Beginning with an intuitive idea of the area under a curve, we shall progress to a deeper understanding of this concept. By considering a very special case at the outset, we can see what is reasonable to try for a definition.

Suppose $f: [a, b] \to R$ is bounded and $f(x) \geq 0$ for all $x \in [a, b]$. For the purpose of visualizing the notion of the area under the curve, it is worthwhile to assume for the moment that f is continuous. We want to determine some reasonable manner of assigning an area to the portion of the plane bounded by the lines $x = a$, $x = b$, $y = 0$, and the graph of f. The approach will be to attempt to approximate the area by use of rectangles, whose areas are easy to compute, and then to use some type of limit process to arrive at our result. Because of the possible pathological nature of f (if we don't assume continuity), it is reasonable to expect some difficulty with this limit idea.

Choose points $x_0, x_1, x_2, \ldots, x_n$ such that $a = x_0 < x_1 < \cdots < x_n = b$. For each i among $1, \ldots, n$, erect a rectangle R_i with the interval $[x_{i-1}, x_i]$ as base and with altitude such that the portion of the graph of f for $x \in [x_{i-1}, x_i]$ lies in this rectangle. This is possible since f is bounded; in fact, we may choose the altitude of R_i to be

$$M_i = \sup \{f(x) : x \in [x_{i-1}, x_i]\}.$$

The region in question lies in the union of the rectangles $R_1, \ldots, R_n$, and these rectangles are nonoverlapping; hence, the area should be less than or equal to the sum of these areas—that is, less than or equal to

$$\sum_{i=1}^{n} M_i(x_i - x_{i-1}).$$

For the more conservative reader, possibly one should try underestimating the area. This time, for each i, erect a rectangle S_i with base $[x_{i-1}, x_i]$ and with altitude such that the rectangle S_i lies entirely inside the region in question. Here we shall choose the altitude of S_i to be

$$m_i = \inf \{f(x) : x \in [x_{i-1}, x_i]\}.$$

The rectangles $S_i, \ldots, S_n$ are nonoverlapping, and all lie inside the portion of the plane bounded by the lines $x = a$, $x = b$, $y = 0$, and the graph of f, hence the sum of their areas,

$$\sum_{i=1}^{n} m_i(x_i - x_{i-1}),$$

must be less than or equal to the sought-after area. See Figure 5.1.

Let us temporarily refer to $\sum_{i=1}^{n} M_i(x_i - x_{i-1})$ as an *upper sum* for f and to $\sum_{i=1}^{n} m_i(x_i - x_{i-1})$ as a *lower sum* for f. The upper sum clearly overestimates the area, and the lower sum underestimates it. Observe that, as one chooses more points in the interval, the upper sum decreases and the lower sum increases; hence they should "converge" to the "area" to be found. The words in quotation marks are used only because these notions have not been formalized in this setting. Let us now formulate a policy concerning the limiting process to which we have made vague reference above. Roughly, we will insist that the upper and lower sums get closer and closer together as we choose more points in $[a, b]$. The discussion just presented is intended to prepare the reader for the precise formulation that is necessary to give the full meaning of the Riemann integral.

5.1 THE RIEMANN INTEGRAL

Consider $a < b$. A *partition* P of $[a, b]$ is any finite set $\{x_0, x_1, \ldots, x_n\}$ such that $a = x_0 < x_1 < \cdots < x_n = b$. If P and Q are partitions of $[a, b]$ with $P \subset Q$, then Q is said to be a *refinement* of P.

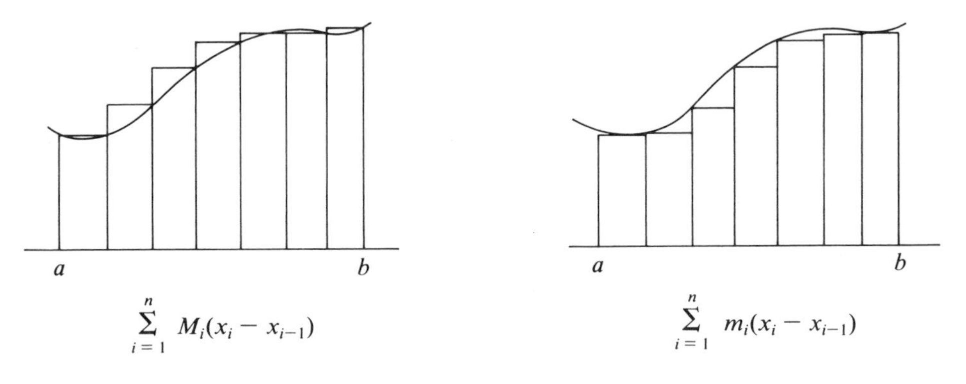

Figure 5.1

Suppose that $f\colon [a, b] \to R$ is a bounded function and

$$P = \{x_0, x_1, \ldots, x_n\}$$

is a partition of $[a, b]$. For each i among $1, \ldots, n$, define

$$M_i(f) = \sup \{f(x) : x \in \{x_{i-1}, x_i]\}$$

and

$$m_i(f) = \inf \{f(x) : x \in \{x_{i-1}, x_i]\}.$$

Define

$$U(P, f) = \sum_{i=1}^{n} M_i(f)(x_i - x_{i-1})$$

and

$$L(P, f) = \sum_{i=1}^{n} m_i(f)(x_i - x_{i-1}).$$

We shall call $U(P, f)$ an *upper sum* for f and $L(P, f)$ a *lower sum* for f. Now f is assumed bounded on $[a, b]$, so there is a real number M such that $-M \leqslant f(x) \leqslant M$ for all $x \in [a, b]$. Thus, in particular, for any partition P of $[a, b]$, $-M(b - a) \leqslant L(P, f) \leqslant U(P, f) \leqslant M(b - a)$. Let

$$\overline{\int_a^b} f\, dx = \inf \{U(P, f) : P \text{ a partition of } [a, b]\}$$

and

$$\underline{\int_a^b} f\, dx = \sup \{L(P, f) : P \text{ a partition of } [a, b]\};$$

in light of the previous sentence, $\overline{\int_a^b} f\, dx$ and $\underline{\int_a^b} f\, dx$, exist and are referred to as the *upper* and *lower integrals* of f, respectively. We say that f is *Riemann-integrable* on $[a, b]$ if and only if

$$\overline{\int_a^b} f\, dx = \underline{\int_a^b} f\, dx,$$

and we define

$$\int_a^b f\, dx = \overline{\int_a^b} f\, dx = \underline{\int_a^b} f\, dx$$

to be the *Riemann integral* of f on $[a, b]$. It will be convenient to write $f \in R(x)$ on $[a, b]$ when f is Riemann-integrable on $[a, b]$. The reader is probably accustomed to writing $\int_a^b f(x)\,dx$ in place of our notation $\int_a^b f\,dx$, and we shall use that notation when convenient. We shall also omit the reference to Riemann from time to time and speak of f being *integrable* on $[a, b]$ and refer to $\int_a^b f\,dx$ as the *integral* of f.

The first theorem to be presented will be an aid in understanding this approach to integration.

5.1 THEOREM Let $f\colon [a, b] \to R$ be bounded. Then if P and Q are any partitions of $[a, b]$, we have

1. If $P \subset Q$, then $L(P, f) \leq L(Q, f)$ and $U(Q, f) \leq U(P, f)$.
2. $L(P, f) \leq U(Q, f)$.
3. $\underline{\int_a^b} f\,dx \leq \overline{\int_a^b} f\,dx$.

Proof.

1. Let $P = \{x_0, x_1, \ldots, x_n\}$, and suppose we add one point z to this partition, $x_{j-1} < z < x_j$; call the new partition P'. Let

$$\begin{aligned}
M_i(f) &= \sup \{f(x) : x \in [x_{i-1}, x_i]\},\\
m_i(f) &= \inf \{f(x) : x \in [x_{i-1}, x_i]\},\\
r_1 &= \sup \{f(x) : x \in \{x_{j-1}, z]\},\\
r_2 &= \sup \{f(x) : x \in [z, x_j]\},\\
s_1 &= \inf \{f(x) : x \in [x_{j-1}, z]\},\\
s_2 &= \inf \{f(x) : x \in [z, x_j]\}.
\end{aligned}$$

Now $M_j(f) \geq \max \{r_1, r_2\}$ and $m_j(f) \leq \min \{s_1, s_2\}$. Then

$$\begin{aligned}
L(P, f) &= \sum_{i=1}^{n} m_i(f)(x_i - x_{i-1})\\
&= \sum_{i=1}^{j-1} m_i(f)(x_i - x_{i-1}) + m_j(f)(x_j - x_{j-1})\\
&\quad + \sum_{i=j+1}^{n} m_i(f)(x_i - x_{i-1})\\
&\leq \sum_{i=1}^{j-1} m_i(f)(x_i - x_{i-1}) + s_1(z - x_{j-1}) + s_2(x_j - z)\\
&\quad + \sum_{i=j+1}^{n} m_i(f)(x_i - x_{i-1})\\
&= L(P', f)
\end{aligned}$$

and

$$\begin{aligned}
U(P, f) &= \sum_{i=1}^{n} M_i(f)(x_i - x_{i-1}) \\
&= \sum_{i=1}^{j-1} M_i(f)(x_i - x_{i-1}) + M_j(f)(x_j - x_{j-1}) \\
&\quad + \sum_{i=j+1}^{n} M_i(f)(x_i - x_{i-1}) \\
&\geqslant \sum_{i=1}^{j-1} M_i(f)(x_i - x_{i-1}) + r_1(z - x_{j-1}) \\
&\quad + r_2(x_j - z) + \sum_{i=j+1}^{n} M_j(f)(x_i - x_{i-1}) \\
&= U(P', f).
\end{aligned}$$

If Q contains k points not in P, repeat this argument k times to obtain $L(P, f) \leqslant L(Q, f)$ and $U(P, f) \geqslant U(Q, f)$.

2. Note first of all that, for any partition P of $[a, b]$, $m_i(f) \leqslant M_i(f)$; hence $L(P, f) \leqslant U(P, f)$. Now, if P and Q are partitions of $[a, b]$, then $P \cup Q$ is also a partition of $[a, b]$, which is a refinement of both P and Q. Then by (1) we have

$$L(P, f) \leqslant L(P \cup Q, f) \leqslant U(P \cup Q, f) \leqslant U(Q, f).$$

3. By (2), $L(P, f) \leqslant U(Q, f)$ for all partitions P and Q of $[a, b]$; hence, for each partition P of $[a, b]$, $L(P, f)$ is a lower bound for $\{U(Q, f) : Q \text{ is a partition of } [a, b]\}$. Thus, $L(P, f) \leqslant \overline{\int_a^b} f\,dx$ for all partitions P; hence

$$\underline{\int_a^b} f\,dx = \sup\{L(P, f) : P \text{ a partition of } [a, b]\} \leqslant \overline{\int_a^b} f\,dx.$$

Consider what is implied by the statement "f is Riemann-integrable on $[a, b]$." First of all, f must be a bounded real-valued function defined on $[a, b]$. Second,

$$\underline{\int_a^b} f\,dx = \overline{\int_a^b} f\,dx.$$

Since $\underline{\int_a^b} f\,dx$ is the supremum of all the lower sums, there must be lower sums very close to $\underline{\int_a^b} f\,dx$. Likewise, since $\overline{\int_a^b} f\,dx$ is the infimum of all the upper sums, there must be upper sums very close to $\overline{\int_a^b} f\,dx$. The integrability of f implies that

$$\overline{\int_a^b} f\,dx = \underline{\int_a^b} f\,dx;$$

hence there must be upper and lower sums close to each other. It is not at all surprising that the converse is also true. Of course, we are referring to a theorem to be stated in in more precise terms than those given in this paragraph.

5.2 THEOREM Let f: $[a, b] \to R$ be bounded. Then $f \in R(x)$ on $[a, b]$ iff for each $\epsilon > 0$ there is a partition P such that

$$U(P, f) - L(P, f) \leq \epsilon.$$

Proof. Suppose $f \in R(x)$ on $[a, b]$. Then

$$\overline{\int_a^b} f\,dx = \underline{\int_a^b} f\,dx = \int_a^b f\,dx.$$

Choose $\epsilon > 0$. Now $\int_a^b f\,dx = \sup\{L(P, f) : P \text{ a partition of } [a, b]\}$, hence $\int_a^b f\,dx - \frac{\epsilon}{2}$ is not an upper bound for the set of all lower sums; so there is a partition P_1 such that

$$\int_a^b f\,dx - \frac{\epsilon}{2} < L(P_1, f).$$

Similarly $\int_a^b f\,dx = \inf\{U(P, f) : P \text{ is a partition of } [a, b]\}$, hence $\int_a^b f\,dx + \frac{\epsilon}{2}$ is not a lower bound for the set of all upper sums; thus, there is a partition P_2 such that

$$U(P_2, f) < \int_a^b f\,dx + \frac{\epsilon}{2}.$$

Let $P = P_1 \cup P_2$. Then

$$\int_a^b f\,dx - \frac{\epsilon}{2} < L(P_1, f) \leq L(P, f) \leq U(P, f)$$
$$\leq U(P_2, f) < \int_a^b f\,dx + \frac{\epsilon}{2}.$$

Thus, $U(P, f) - L(P, f) \leq \epsilon$.

Suppose $f\colon [a, b] \to R$ is bounded and such that for each $\epsilon > 0$ there is a partition P such that

$$U(P, f) - L(P, f) \leq \epsilon.$$

We need to show that

$$\overline{\int_a^b} f\,dx = \underline{\int_a^b} f\,dx.$$

By Theorem 5.1, $\overline{\int_a^b} f\,dx - \underline{\int_a^b} f\,dx \geq 0$. We shall show that for each $\epsilon > 0$,

$$0 \leq \overline{\int_a^b} f\,dx - \underline{\int_a^b} f\,dx \leq \epsilon,$$

hence

$$\overline{\int_a^b} f\,dx = \underline{\int_a^b} f\,dx.$$

Choose $\epsilon > 0$. There is a partition P such that

$$U(P, f) - L(P, f) \leq \epsilon.$$

Thus, we have

$$L(P, f) \leq \underline{\int_a^b} f\,dx \leq \overline{\int_a^b} f\,dx \leq U(P, f) \leq L(P, f) + \epsilon.$$

Therefore, $0 \leq \overline{\int_a^b} f\,dx - \underline{\int_a^b} f\,dx \leq \epsilon$; hence

$$\overline{\int_a^b} f\,dx = \underline{\int_a^b} f\,dx$$

and $f \in R(x)$ on $[a, b]$.

Consider the geometric interpretation of $U(P, f) - L(P, f)$ for the case in which $f(x) \geq 0$ for all x. In Figure 5.2, the area of the shaded region is $U(P, f) - L(P, f)$. From this point of view, Theorem 5.2 is very natural. The situation that arises here is not uncommon. Theorem 5.2 gives a condition for integrability, but it does not give us any simple means for computing the integral. We encountered the same type of problem in our discussion of sequences.

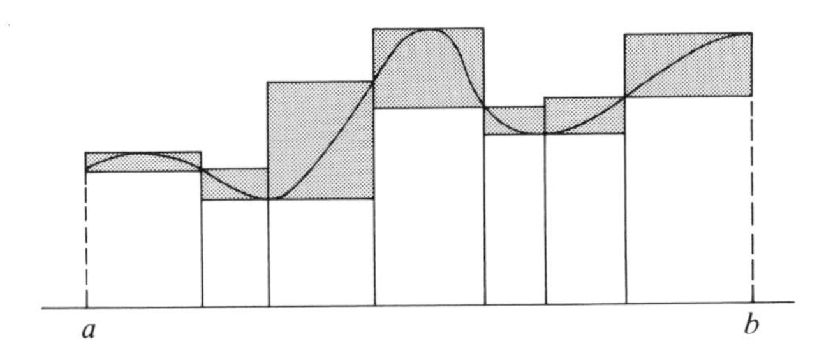

Figure 5.2

■ **Example 5.1** Let us consider a few examples now. Consider the function $f: [0, 1] \to R$ such that $f(x) = 0$ if x is rational and $f(x) = 1$ if x is irrational. Let P be any partition of $[a, b]$, $P = \{x_0, x_1, \ldots, x_n\}$. Since every interval $[x_{i-1}, x_i]$ contains both rational and irrational points, $M_i(f) = 1$ and $m_i(f) = 0$. Hence, $U(P, f) = 1$ and $L(P, f) = 0$. Since all upper sums are equal to 1 and all lower sums are equal to 0, we have

$$\overline{\int_0^1} f\,dx = 1 \quad \text{and} \quad \underline{\int_0^1} f\,dx = 0$$

so f is not Riemann-integrable on $[0, 1]$. ■

■ **Example 5.2** Consider now the function $f: [0, 1] \to R$ such that $f(x) = x^2$ for all $x \in [0, 1]$. Choose $\epsilon > 0$, and let $P = \{x_0, x_1, \ldots, x_n\}$ be any partition of $[0, 1]$ such that

$$\max \{x_i - x_{i-1} : i = 1, \ldots, n\} < \frac{\epsilon}{2}.$$

Since f is increasing and continuous,

$$M_i(f) = f(x_i) = x_i^2 \quad \text{and} \quad m_i(f) = f(x_{i-1}) = x_{i-1}^2;$$

hence

$$U(P, f) = \sum_{i=1}^{n} x_i^2\,(x_i - x_{i-1}) \quad \text{and} \quad L(P, f) = \sum_{i=1}^{n} x_{i-1}^2\,(x_i - x_{i-1}).$$

Thus, we have

$$\begin{aligned} U(P, f) - L(P, f) &= \sum_{i=1}^{n} (x_i^2 - x_{i-1}^2)(x_i - x_{i-1}) \\ &= \sum_{i=1}^{n} (x_i + x_{i-1})(x_i - x_{i-1})(x_i - x_{i-1}) \\ &< \sum_{i=1}^{n} 2\left(\frac{\epsilon}{2}\right)(x_i - x_{i-1}) \end{aligned}$$

$$= \epsilon \sum_{i=1}^{n} (x_i - x_{i-1}) = \epsilon.$$

This means that f is Riemann-integrable on [0, 1]. Note again that we have no information yet as to the value of $\int_0^1 x^2\, dx$. We shall attack that problem soon.

■

■ **Example 5.3** Define f: $[0, 1] \to R$ by $f(x) = 0$ if x is irrational and $f(\frac{p}{q}) = \frac{1}{q}$ where p and q are relatively prime integers with $q > 0$. [We define $f(0) = 1$.] Now since every interval contains irrational points, $L(P, f) = 0$ for all partitions P; hence

$$\underline{\int_a^b} f\, dx = 0.$$

If we can show that f is integrable on [0, 1], then we will know

$$\int_a^b f\, dx = \underline{\int_a^b} f\, dx = 0.$$

Choose $0 < \epsilon < 1$. Since $L(P, f) = 0$ for all partitions P, it suffices to find a partition Q such that $U(Q, f) \leqslant \epsilon$. Let $\{r_0, \ldots, r_n\}$ be the set of all rational points in [0, 1] such that $f(r_i) \geqslant \frac{\epsilon}{2}$, which means all rational points in [0, 1] of the form $\frac{p}{q}$ where p and q are relatively prime and $0 < q < \frac{2}{\epsilon}$. We may assume $0 = r_0 < r_1 < \cdots < r_n = 1$. Let us now proceed to construct a partition P such that $U(P, f) < \epsilon$. Define $x_0 = r_0 = 0$, and choose x_1 such that

$$0 < x_1 < \frac{r_1}{2} \quad \text{and} \quad x_1 < \frac{\epsilon}{2(n + 1)}.$$

Choose x_2 and x_3 such that

$$x_1 < x_2 < r_1 < x_3 < \frac{r_1 + r_2}{2} \quad \text{and} \quad x_3 - x_2 < \frac{\epsilon}{2(n + 1)}.$$

Continuing in this fashion (by induction), we obtain $x_0 < x_1 < x_2 < \cdots < x_{2n-1}$ such that

$$x_{2k} < r_k < x_{2k+1} \quad \text{and} \quad x_{2k+1} - x_{2k} < \frac{\epsilon}{2(n + 1)}$$

for $k = 1, 2, \ldots, n - 1$. Choose x_{2n} such that

$$x_{2n-1} < x_{2n} < 1 \quad \text{and} \quad 1 - x_{2n} < \frac{\epsilon}{2(n + 1)}.$$

Let $x_{2n+1} = 1$. This process may seem mysterious, but the key to these choices is the desire to guarantee that, where the function takes values larger than $\frac{\epsilon}{2}$, the width of the rectangle is small. Now if we let $P = \{x_0, x_1, \ldots, x_{2n+1}\}$, we have

$$U(P, f) = \sum_{i=1}^{2n+1} M_i(f)(x_i - x_{i-1}) = \sum_{k=0}^{n} M_{2k+1}(f)(x_{2k+1} - x_{2k})$$

$$+ \sum_{k=1}^{n} M_{2k}(f)(x_{2k} - x_{2k-1}) < \sum_{k=0}^{n} (x_{2k+1} - x_{2k})$$

$$+ \sum_{k=1}^{n} \frac{\epsilon}{2}(x_{2k} - x_{2k-1}) < (n + 1)\frac{\epsilon}{2(n + 1)} + \frac{\epsilon}{2} = \epsilon.$$

Thus, $f \in R(x)$ on $[0, 1]$ and $\int_0^1 f\, dx = 0$. ■

In this example, we see a departure from the idea of the integral being the area under a curve. In this case, the function f is nonnegative and takes positive values at every rational point but $\int_0^1 f\, dx = 0$.

5.2 CLASSES OF INTEGRABLE FUNCTIONS

There are some obvious classes of functions that one might suspect are integrable—the class of continuous functions, for example. Before taking up that problem, let us consider the class of monotone functions.

5.3 THEOREM If f: $[a, b] \to R$ is monotone, then $f \in R(x)$ on $[a, b]$.

Proof. Since f is monotone, f is bounded on $[a, b]$. Let us assume f is increasing. The case for f decreasing is similar. Choose $\epsilon > 0$. There is $k > 0$ such that

$$k[(f(b) - f(a)] < \epsilon.$$

Choose a partition $P = \{x_0, x_1, \ldots, x_n\}$ such that

$$x_i - x_{i-1} < k$$

for $i = 1, 2, \ldots, n$. Since f is *increasing*, $M_i(f) = f(x_i)$ and $m_i(f) = f(x_{i-1})$. Now

$$U(P, f) - L(P, f) = \sum_{i=1}^{n} [f(x_i) - f(x_{i-1})][x_i - x_{i-1}]$$

$$\leqslant \sum_{i=1}^{n} [f(x_i) - f(x_{i-1})]k = k[f(b) - f(a)] < \epsilon.$$

Thus, $f \in R(x)$ on $[a, b]$.

Notice that Theorem 5.3 handles the example $f(x) = x^2$ for $x \in [0, 1]$, which we discussed before.

To enlarge the class of integrable functions, let us consider the continuous functions. Let $f: [a, b] \to R$ be continuous. Since $[a, b]$ is compact, f is bounded, so we need only consider $U(P, f) - L(P, f)$ and try to make this difference small. If

$$P = \{x_0, x_1, \ldots, x_n\},$$

then by the continuity of f there are $t_i, s_i \in [x_{i-1}, x_i]$ for $i = 1, 2, \ldots, n$ such that

$$M_i(f) = f(t_i) \quad \text{and} \quad m_i(f) = f(s_i).$$

Now

$$\begin{aligned} U(P, f) - L(P, f) &= \sum_{i=1}^{n} M_i(f)(x_i - x_{i-1}) \\ &\quad - \sum_{i=1}^{n} m_i(f)(x_i - x_{i-1}) \\ &= \sum_{i=1}^{n} [f(t_i) - f(s_i)](x_i - x_{i-1}). \end{aligned}$$

If we force $f(t_i) - f(s_i) < \gamma$, then

$$U(P, f) - L(P, f) < \gamma(b - a).$$

Of course, here is where the continuity of f plays its role. Lest we give away all secrets in advance, let us state and prove the obvious theorem.

5.4 THEOREM If $f: [a, b] \to R$ is continuous, then $f \in R(x)$ on $[a, b]$.

Proof. As observed above, f is bounded. Choose $\epsilon > 0$. Since $[a, b]$ is compact, f is uniformly continuous on $[a, b]$, so there is $\delta > 0$ such that $|x - y| < \delta$ and $x, y \in [a, b]$ implies

$$|f(x) - f(y)| < \frac{\epsilon}{b - a}.$$

Let $P = \{x_0, x_1, \ldots, x_n\}$ be any partition such that $x_i - x_{i-1} < \delta$ for $i = 1, \ldots, n$. There are $t_i, s_i \in [x_{i-1}, x_i]$ such that $M_i(f) = f(t_i)$ and $m_i(f) = f(s_i)$ for $i = 1, 2, \ldots, n$. Since $x_i - x_{i-1} < \delta$, $|t_i - s_i| < \delta$, and hence

$$0 \leqslant f(t_i) - f(s_i) < \frac{\epsilon}{b - a}.$$

Thus, we have

$$U(P, f) - L(P, f) = \sum_{i=1}^{n} [f(t_i) - f(s_i)](x_i - x_{i-1})$$

$$\leq \sum_{i=1}^{n} \frac{\epsilon}{b - a}(x_i - x_{i-1})$$

$$= \frac{\epsilon}{b - a}(b - a) = \epsilon.$$

Therefore, $f \in R(x)$ on $[a, b]$.

In the discussion at the beginning of the chapter, the approach to the integral involved only upper and lower sums or, for nonnegative functions, overestimates and underestimates of the area under the curve. Let us now take a more liberal point of view. Suppose that f: $[a, b] \to R$ is bounded and $P = \{x_0, x_1, x_2, \ldots, x_n\}$ is a partition of $[a, b]$. For each i among $1, \ldots, n$, choose $t_i \in [x_{i-1}, x_n\}$ and form the sum

$$\sum_{i=1}^{n} f(t_i)(x_i - x_{i-1}).$$

Now for each i among $1, \ldots, n$,

$$m_i(f) \leq f(t_i) \leq M_i(f);$$

hence

$$L(P, f) \leq \sum_{i=1}^{n} f(t_i)(x_i - x_{i-1}) \leq U(P, f).$$

Since for integrable functions, $L(P, f)$ and $U(P, f)$ squeeze in on $\int_a^b f\,dx$, it seems reasonable to suspect that the new sums just considered behave similarly. Indeed, as we have seen, for continuous functions, each upper and lower sum can be written in this way. If this were the case, possibly wise choices for the points t_i might be useful in determining $\int_a^b f\,dx$. Let us illustrate this point with a simple example.

■ **Example 5.4** We have seen that the function $f(x) = x^2$ is integrable on the interval $[0, 1]$. Let $P = \{x_0, x_1, \ldots, x_n\}$ be any partition of $[0, 1]$. Consider the function $g(x) = \frac{x^3}{3}$. (Are you surprised by this choice?) Now g is continuous and differentiable on $[0, 1]$ and $g'(x) = f(x)$ for all $x \in [0, 1]$. For each i among $1, \ldots, n$, the Mean-Value Theorem applied to g on $[x_{i-1}, x_i]$ yields $t_i \in [x_{i-1}, x_i]$ such that

$$g(x_i) - g(x_{i-1}) = g'(t_i)(x_i - x_{i-1}) = f(t_i)(x_i - x_{i-1}).$$

Hence,

$$\sum_{i=1}^{n} f(t_i)(x_i - x_{i-1}) = \sum_{i=1}^{n} [g(x_i) - g(x_{i-1})] = g(1) - g(0) = \frac{1}{3}.$$

By our remarks in the preceding paragraph, $L(P, f) \leq \frac{1}{3} \leq U(P, f)$ for all partitions P; hence

$$\underline{\int_a^b} f\,dx \leq \frac{1}{3} \leq \overline{\int_a^b} f\,dx \quad \text{or} \quad \int_a^b x^2\,dx = \frac{1}{3}.$$

This result is not surprising. The basic idea involved will come later. ■

5.3 RIEMANN SUMS

Lest we stray too far afield, let us attempt to formulate a condition for integrability involving the type of sum mentioned above. Let P be any partition of $[a, b]$, $P = \{x_0, x_1, \ldots, x_n\}$. Define the *mesh* of P to be the maximum of the lengths $x_i - x_{i-1}$, $i = 1, \ldots, n$, and denote it by $\mu(P)$. If $t_i \in [x_{i-1}, x_i]$ are chosen for $i = 1, \ldots, n$, P is called a *marked* partition, and

$$S(P, f) = \sum_{i=1}^{n} f(t_i)(x_i - x_{i-1})$$

is called a *Riemann sum* for f. There is some ambiguity of notation involved since a partition P may be marked in many ways, but we shall try to handle this problem in a manner that avoids any confusion.

5.5 THEOREM Let f: $[a, b] \to R$ be bounded. Then $f \in R(x)$ on $[a, b]$ iff there is a real number A such that for each $\epsilon > 0$ there is a partition P such that for any refinement Q of P, regardless of how marked, $|S(Q, f) - A| \leq \epsilon$. If the latter condition is satisfied, then

$$A = \int_a^b f\,dx.$$

Proof. Suppose $f \in R(x)$ on $[a, b]$. Let $A = \int_a^b f\,dx$ and choose $\epsilon > 0$. There is a partition P such that $U(P, f) - L(P, f) \leq \epsilon$. Let Q be any refinement of P marked in any fashion. Then

$$L(P, f) \leq L(Q, f) \leq S(Q, f) \leq U(Q, f) \leq U(P, f) \leq L(P, f) + \epsilon,$$

and

$$L(Q, f) \leq \int_a^b f\,dx \leq U(Q, f).$$

Hence

$$|S(Q, f) - \int_a^b f\,dx| \leq \epsilon.$$

Suppose $f: [a, b] \to R$ is bounded and A is a real number such that for each $\epsilon > 0$ there is a partition P such that if $P \subset Q$, where Q is a partition of $[a, b]$ marked in any fashion, then $|S(Q, f) - A| \leqslant \epsilon$. We shall appeal to Theorem 5.2. Choose $\epsilon > 0$. There is a partition $P = \{x_0, x_1, \ldots, x_n\}$ such that

$$|S(P, f) - A| < \frac{\epsilon}{4},$$

regardless of how we choose $t_i \in [x_{i-1}, x_i]$. Recall that

$$M_i(f) = \sup \{f(x) : x \in [x_{i-1}, x_i]\}$$

and

$$m_i(f) = \inf \{f(x) : x \in [x_{i-1}, x_i]\}.$$

Thus, for each i among $1, \ldots, n$, there are $t_i, s_i \in [x_{i-1}, x_i]$ such that

$$M_i(f) - \frac{\epsilon}{4(b - a)} < f(t_i)$$

and

$$f(s_i) < m_i(f) + \frac{\epsilon}{4(b - a)}.$$

Now

$$\begin{aligned}
U(P, f) - L(P, f) &= \sum_{i=1}^{n} M_i(f)(x_i - x_{i-1}) \\
&\quad - \sum_{i=1}^{n} m_i(f)(x_i - x_{i-1}) \\
&\leqslant \sum_{i=1}^{n} \left[f(t_i) + \frac{\epsilon}{4(b - a)}\right](x_i - x_{i-1}) \\
&\quad - \sum_{i=1}^{n} \left[f(s_i) - \frac{\epsilon}{4(b - a)}\right](x_i - x_{i-1}) \\
&= \sum_{i=1}^{n} f(t_i)(x_i - x_{i-1}) + \frac{\epsilon}{4} \\
&\quad - \sum_{i=1}^{n} f(s_i)(x_i - x_{i-1}) + \frac{\epsilon}{4} \\
&= \left[\sum_{i=1}^{n} f(t_i)(x_i - x_{i-1}) - A\right] \\
&\quad - \left[\sum_{i=1}^{n} f(s_i)(x_i - x_{i-1}) - A\right] + \frac{\epsilon}{2} \\
&\leqslant \frac{\epsilon}{4} + \frac{\epsilon}{4} + \frac{\epsilon}{2} = \epsilon.
\end{aligned}$$

Thus, $f \in R(x)$ on $[a, b]$. It now remains to show that $A = \int_a^b f\,dx$. Choose $\epsilon > 0$. There is a partition P_1 on $[a, b]$ such that for any refinement Q of P_1, regardless of how marked,

$$|S(Q, f) - A| \leq \frac{\epsilon}{3}.$$

Since $f \in R(x)$, there is a partition P_2 of $[a, b]$ such that

$$U(P_2, f) - L(P_2, f) \leq \frac{\epsilon}{3}.$$

Let $Q = P_1 \cup P_2$. Then

$$\left|A - \int_a^b f\,dx\right| \leq |A - S(Q, f)| + |S(Q, f) - L(Q, f)| + \left|L(Q, f) - \int_a^b f\,dx\right| \leq \frac{\epsilon}{3} + \frac{\epsilon}{3} + \frac{\epsilon}{3} = \epsilon,$$

since

$$L(Q, f) \leq S(Q, f) \leq U(Q, f) \leq L(Q, f) + \frac{\epsilon}{3}$$

and

$$L(Q, f) \leq \int_a^b f\,dx \leq U(Q, f) \leq L(Q, f) + \frac{\epsilon}{3}.$$

Since $\epsilon > 0$ was arbitrary, $A = \int_a^b f\,dx$.

To show that a function is integrable in the examples we have considered, it actually has sufficed to find a partition with sufficiently small mesh. This is not accidental, as can be seem from the next theorem.

5.6 THEOREM Suppose f: $[a, b] \to R$ is bounded. Then $f \in R(x)$ on $[a, b]$ iff there is a number A such that for each $\epsilon > 0$ there is $\delta > 0$ such that for any partition P of $[a, b]$ with $\mu(P) < \delta$, regardless of how P is marked,

$$|S(P, f) - A| \leq \epsilon.$$

If the condition is satisfied, then $A = \int_a^b f\,dx$.

Proof. Suppose that f: $[a, b] \to R$ is bounded and that there is a real number A such that for any $\epsilon > 0$, there is $\delta > 0$ such that for any partition P

of $[a, b]$ with $\mu(P) < \delta$, we have $|S(P, f) - A| \leq \epsilon$, regardless of how P is marked. Then, if P is any partition of $[a, b]$ with $\mu(P) < \delta$ and Q is a refinement of P, then $\mu(Q) < \delta$, and hence, regardless of how Q is marked, $|S(Q, f) - A| \leq \epsilon$. By Theorem 5.5, then, $f \in R(x)$ and $A = \int_a^b f\,dx$.

Suppose now that $f \in R(x)$ on $[a, b]$. Choose $\epsilon > 0$. Let $A = \int_a^b f\,dx$ with M a positive real number such that $|f(x)| \leq M$ for all $x \in [a, b]$ and $\epsilon' = \frac{\epsilon}{4}$. There is a partition $P = \{x_0, x_1, \ldots, x_n\}$ such that

$$U(P, f) - L(P, f) < \epsilon'.$$

Let

$$\delta = \frac{\epsilon}{16Mn}.$$

Let $P' = \{z_0, z_1, \ldots, z_r\}$ be any partition of $[a, b]$ such that $\mu(P') < \delta$. Consider any interval $[z_{i-1}, z_i]$, and suppose

$$\{x_j, x_{j+1}, \ldots, x_{j+k}\} \subset [z_{i-1}, z_i].$$

Let

$$\begin{aligned} M_k(f) &= \sup\{f(x) : x \in [z_{k-1}, z_k]\}, \\ m_k(f) &= \inf\{f(x) : x \in [z_{k-1}, z_k]\}, \\ M'_k(f) &= \sup\{f(x) : x \in [x_{k-1}, x_k]\}, \quad \text{and} \\ m'_k(f) &= \inf\{f(x) : x \in [x_{k-1}, x_k]\}. \end{aligned}$$

Now

$$M_i(f) \leq 2M + M'_{j+k}(f)$$

for $s = 1, \ldots, k,$

$$M_i(f) \leq 2M + \sup\{f(x) : x \in [z_{i-1}, x_j]\},$$

and

$$M_i(f) \leq 2M + \sup\{f(x) : x \in [x_{j+k}, z_i]\}.$$

Similarly,

$$m_i(f) \geq -2M + m'_{j+s}(f)$$

for $s = 1, \ldots, k,$

$$m_i(f) \geq -2M + \inf\{f(x) : x \in [z_{i-1}, x_j]\},$$

and

$$m_i(f) \geq -2M + \inf\{f(x) : x \in [x_{j+k}, z_i]\}.$$

Now

$$U(P', f) = \sum_{k=1}^{r} M_k(f)(z_k - z_{k-1}) \leq U(P \cup P', f) + 2M(2n)\delta$$

and

$$L(P', f) = \sum_{k=1}^{r} m_k(f)(z_k - z_{k-1}) \geq L(P \cup P', f) - 2M(2n)\delta.$$

Choose any $t_i \in [z_{i-1}, z_i]$ for $i = 1, \ldots, r$. Then

$$\begin{aligned} L(P, f) - \frac{\epsilon}{2} &\leq L(P, f) - 4Mn\delta \\ &\leq L(P \cup P', f) - 4Mn\delta \leq L(P', f) \leq A \end{aligned}$$

and

$$\begin{aligned} A \leq U(P', f) &\leq U(P \cup P', f) + 4Mn\delta \\ &\leq U(P, f) + \frac{\epsilon}{4} \leq L(P, f) + \frac{\epsilon}{2}. \end{aligned}$$

Moreover,

$$\begin{aligned} L(P, f) - \frac{\epsilon}{2} \leq L(P', f) &\leq \sum_{i=1}^{r} f(t_i)(z_i - z_{i-1}) \\ &\leq U(P', f) \leq L(P, f) + \frac{\epsilon}{2}. \end{aligned}$$

Hence $\left|\sum_{i=1}^{r} f(t_i)(z_i - z_{i-1}) - A\right| \leq \epsilon$.

In light of our experience with limits and sequences the following theorem seems very natural.

5.7 THEOREM Suppose f: $[a, b] \to R$ is bounded. Then $f \in R(x)$ on $[a, b]$ iff, for each sequence $\{P_n\}_{n=1}^{\infty}$ of marked partitions with $\{\mu(P_n)\}_{n=1}^{\infty}$ converging to zero, the sequence $\{S(P_n, f)\}_{n=1}^{\infty}$ is convergent. If the condition is satisfied then each of the sequences $\{S(P_n, f)\}_{n=1}^{\infty}$ will converge to $\int_a^b f\, dx$.

The proof of this theorem is left to the reader (see Exercise 10) because of its similarity to Theorem 2.1. If the task is difficult, the reader has failed to comprehend the strategy of the proof of Theorem 2.1.

■ **Example 5.5** We will use Theorem 5.7 to establish the identity

$$\int_a^1 \frac{1}{x^2 + 1}\, dx = \int_1^{1/a} \frac{1}{x^2 + 1}\, dx \quad \text{for } a > 0.$$

The function $f(x) = \dfrac{1}{x^2 + 1}$ is continuous on R, so it is integrable on every closed interval. Choose a sequence $\{P_n\}_{n=1}^{\infty}$ of partitions of $[a, 1]$ with $\{\mu(P_n\}_{n=1}^{\infty}$ converging to zero. If $P_n = \{x_0, x_1, \ldots, x_m\}$, choose $t_i \in [x_{i-1}, x_i]$ such that $t_i^2 = x_{i-1}x_i$; that is, t_i is the geometric mean of x_{i-1} and x_i. With each partition so marked, $\{S(P_n, f)\}_{n=1}^{\infty}$ converges to $\displaystyle\int_a^1 \frac{1}{x^2+1}\,dx$. Define a sequence of partitions of $\left[1, \dfrac{1}{a}\right]$ as follows: For each $P_n = \{x_0, x_1, x_2, \ldots, x_m\}$, define $P'_n = \left\{\dfrac{1}{x_m}, \dfrac{1}{x_{m-1}}, \ldots, \dfrac{1}{x_1}, \dfrac{1}{x_0}\right\}$ and mark P'_n with $t'_i = \dfrac{1}{t_i}$ where the t_i are chosen as indicated above. Since P_n is a partition of $[a, 1]$, P'_n is a partition of $\left[1, \dfrac{1}{a}\right]$ and $\mu(P'_n) \leq \dfrac{1}{a^2}\mu(P_n)$; hence $\{\mu(P'_n)\}_{n=1}^{\infty}$ converges to 0. Thus $\{S(P'_n, f)\}_{n=1}^{\infty}$ converges to $\displaystyle\int_1^{1/a} \frac{1}{x^2+1}\,dx$. But $S(P_n, f) = S(P'_n, f)$ for each n, hence $\displaystyle\int_a^1 \frac{1}{x^2+1}\,dx = \int_1^{1/a} \frac{1}{x^2+1}\,dx.$ ■

5.4 THE FUNDAMENTAL THEOREM OF INTEGRAL CALCULUS

It is impossible to resist the temptation to use one of these results to prove a very important theorem, the Fundamental Theorem of Integral Calculus. This is indeed a rather impressive result in the sense that it gives a method for computing certain integrals and reveals the relationship between integration and differentiation. We have already revealed the secret of the proof to be used here in the discussion preceding Theorem 5.5.

5.8 FUNDAMENTAL THEOREM OF INTEGRAL CALCULUS Suppose $f\colon [a, b] \to R$ is differentiable on $[a, b]$ and $f' \in R(x)$ on $[a, b]$. Then

$$\int_a^b f'\,dx = f(b) - f(a).$$

Proof. Let P be any partition of $[a, b]$, $P = \{x_0, x_1, \ldots, x_n\}$. By the Mean-Value Theorem applied to f on $[x_{i-1}, x_i]$, there is $t_i \in [x_{i-1}, x_i]$ such that

$$f(x_i) - f(x_{i-1}) = f'(t_i)(x_i - x_{i-1}).$$

Thus,

$$\sum_{i=1}^{n} f'(t_i)(x_i - x_{i-1}) = \sum_{i=1}^{n} [f(x_i) - f(x_{i-1})] = f(b) - f(a).$$

This means that any partition P of $[a, b]$ may be marked in a way so that $S(P, f') = f(b) - f(a)$. Choose a sequence $\{P_n\}_{n=1}^{\infty}$ of partitions of $[a, b]$, each marked in this fashion and such that $\{\mu(P_n)\}_{n=1}^{\infty}$ converges to zero. Now, $f' \in R(x)$ on $[a, b]$; hence, by Theorem 5.7, $S(P_n, f')$ converges to $\int_a^b f' \, dx$, but

$$S(P_n, f') = f(b) - f(a)$$

for each n, so $\int_a^b f' \, dx = f(b) - f(a)$.

It should be pointed out that $f' \in R(x)$ is a very important part of the hypothesis of this theorem. For example, the function f defined on $[0, 1]$ by $f(x) = x^2 \sin \frac{1}{x^2}$ for $x \neq 0$ and $f(0) = 0$ is differentiable on $[0, 1]$, but f' is not Riemann-integrable on $[0, 1]$ since it is unbounded.

The reader will no doubt recognize this theorem as the basis for many calculations of integrals in early calculus courses. The next example is intended to reinforce that recognition.

■ **Example 5.6** We will use Theorem 5.8 to calculate the integral of $F(x) = x^3 - 1$ over the interval $[2, 3]$. First of all, F is integrable because it is continuous. So we may utilize Theorem 5.8 if we can find a function f such that $f'(x) = F(x)$ for all $x \in [2, 3]$—that is, find an antiderivative of F. Experience helps us discover that $f(x) = \frac{x^4}{4} - x$ serves our purpose. Then

$$\int_2^3 (x^3 - 1) \, dx = \int_2^3 F(x) \, dx = \int_2^3 f'(x) \, dx = f(3) - f(2) = \frac{61}{4}. \quad ■$$

Theorem 5.8 does have some limitations. If we wish to use it to find the integral of F, then we must find an antiderivative of F—that is, a function f such that $f'(x) = F(x)$. Under some circumstances, this may be very difficult to do; you may recall quite a bit of time devoted to this problem in your calculus courses. But now we know that it may be impossible! In order for F to have an antiderivative, it must be the derivative of f and, hence, must have the intermediate-value property (see Theorem 4.11). If F fails to have the intermediate-value property, it cannot have an antiderivative; hence, Theorem 5.8 is of no use in finding the integral of F.

5.5 ALGEBRA OF INTEGRABLE FUNCTIONS

It is now appropriate to make some observations useful in calculating integrals. Since the integral may be considered as a limit of sums in a certain way, the results are derived from similar results for finite sums.

5.9 THEOREM If $f_1, f_2: [a, b] \to R$ are bounded and $f_1, f_2 \in R(x)$ on $[a, b]$, then

a. For c_1, c_2 any real numbers, $c_1 f_1 + c_2 f_2 \in R(x)$ on $[a, b]$ and

$$\int_a^b (c_1 f_1 + c_2 f_2)\, dx = c_1 \int_a^b f_1\, dx + c_2 \int_a^b f_2\, dx.$$

b. If $f_1(x) \leqslant f_2(x)$ for all $x \in [a, b]$, then

$$\int_a^b f_1(x)\, dx \leqslant \int_a^b f_2(x)\, dx.$$

c. If $m \leqslant f_1(x) \leqslant M$ for all $x \in [a, b]$, then

$$m(b - a) \leqslant \int_a^b f_1\, dx \leqslant M(b - a).$$

Proof.

a. Choose $\epsilon > 0$. There is $\epsilon' > 0$ such that

$$(|c_1| + |c_2|)\epsilon' \leqslant \epsilon.$$

There are partitions P_1 and P_2 such that for any refinement Q_i of P_i, regardless of how Q_i is marked,

$$\left| S(Q_i, f_i) - \int_a^b f_i\, dx \right| \leqslant \epsilon'.$$

Let $P = P_1 \cup P_2$. If Q is any refinement of P, then Q is a refinement of P_1 and a refinement of P_2; hence

$$\begin{aligned}
&\left| S(Q, c_1 f_1 + c_2 f_2) - \left[c_1 \int_a^b f_1\, dx + c_2 \int_a^b f_2\, dx \right] \right| \\
&\quad = \left| c_1 S(Q, f_1) + c_2 S(Q, f_2) - c_1 \int_a^b f_1\, dx - c_2 \int_a^b f_2\, dx \right| \\
&\quad \leqslant |c_1| \left| S(Q, f_1) - \int_a^b f_1\, dx \right| + |c_2| \left| S(Q, f_2) - \int_a^b f_2\, dx \right| \\
&\quad \leqslant |c_1|\epsilon' + |c_2|\epsilon' \leqslant \epsilon.
\end{aligned}$$

Hence, by Theorem 5.5, $c_1 f_1 + c_2 f_2 \in R(x)$ on $[a, b]$ and

$$\int_a^b (c_1 f_1 + c_2 f_2) dx = c_1 \int_a^b f_1\, dx + c_2 \int_a^b f_2\, dx.$$

b. If $f_1(x) \leqslant f_2(x)$ for all $x \in [a, b]$, then $L(P, f_1) \leqslant L(P, f_2)$ for all partitions P; hence

$$\int_a^b f_1\, dx = \underline{\int_a^b} f_1\, dx \leqslant \underline{\int_a^b} f_2\, dx = \int_a^b f_2\, dx.$$

c. If $m \leqslant f_1(x) \leqslant M$ for all $x \in [a, b]$, then, for any partition P of $[a, b]$,

$$m(b - a) \leqslant L(P, f_1) \leqslant \int_a^b f_1 \, dx \leqslant U(P, f_1) \leqslant M[b - a].$$

5.10 THEOREM Assume that f: $[a, b] \to R$ is bounded. If $a < c < b$, then $f \in R(x)$ on $[a, b]$ iff $f \in R(x)$ on $[a, c]$ and $f \in R(x)$ on $[c, b]$. If $f \in R(x)$ on $[a, b]$, then

$$\int_a^b f \, dx = \int_a^c f \, dx + \int_c^b f \, dx.$$

Proof. Assume $f \in R(x)$ on $[a, c]$ and $f \in R(x)$ on $[c, b]$. Choose $\epsilon > 0$. There are partitions P_1 and P_2 of $[a, c]$ and $[c, b]$, respectively, such that if Q_1 and Q_2 are refinements of P_1 and P_2, respectively, then

$$\left| S(Q_1, f) - \int_a^c f \, dx \right| \leqslant \frac{\epsilon}{2} \quad \text{and} \quad \left| S(Q_2, f) - \int_c^b f \, dx \right| \leqslant \frac{\epsilon}{2}.$$

Let $P = P_1 \cup P_2$. Then P is a partition of $[a, b]$, and if Q is any refinement of P, then $Q_1 = Q \cap [a, c]$ is a refinement of P_1 and $Q_1 = Q \cap [c, b]$ is a refinement of P_2. Thus,

$$\begin{aligned} &\left| S(Q, f) - \left[\int_a^c f \, dx + \int_c^b f \, dx \right] \right| \\ &\quad = \left| S(Q_1, f) + S(Q_2, f) - \int_a^c f \, dx - \int_c^b f \, dx \right| \\ &\quad \leqslant \left| S(Q_1, f) - \int_a^c f \, dx \right| + \left| S(Q_2, f) - \int_c^b f \, dx \right| \leqslant \frac{\epsilon}{2} + \frac{\epsilon}{2} = \epsilon. \end{aligned}$$

Thus, by Theorem 5.5, $f \in R(x)$ on $[a, b]$ and

$$\int_a^b f \, dx = \int_a^c f \, dx + \int_c^b f \, dx.$$

Assume $f \in R(x)$ on $[a, b]$. Choose $\epsilon > 0$. There is a partition P of $[a, b]$ such that

$$U(P, f) - L(P, f) < \epsilon.$$

Let $Q = P \cap \{c\}$ and let $Q_1 = Q \cap [a, c]$ and $Q_2 = Q \cap [c, b]$. Then

$$\begin{aligned} \epsilon &> U(Q, f) - L(Q, f) \\ &= U(Q_1, f) + U(Q_2, f) - [L(Q_1, f) + L(Q_2, f)] \\ &= [U(Q_1, f) - L(Q_1, f)] + [U(Q_2, f) - L(Q_2, f)]. \end{aligned}$$

Since $U(Q_1, f) - L(Q_1, f) > 0$ and $U(Q_2, f) - L(Q_2, f) > 0$, we may conclude

$$U(Q_1, f) - L(Q_1, f) < \epsilon \quad \text{and} \quad U(Q_2, f) - L(Q_2, f) < \epsilon.$$

Thus, by Theorem 5.2, $f \in R(x)$ on $[a, c]$ and $f \in R(x)$ on $[c, b]$.

We have seen that linear combinations of integrable functions are integrable. Now we seek other operations that preserve integrable functions. As has been shown already, continuous functions seem to behave especially well with respect to integration. The next theorem should help further this opinion.

5.11 THEOREM Suppose f: $[a, b] \to S$, $f \in R(x)$ on $[a, b]$, and ϕ: $S \to R$ is continuous with S compact. Then $\phi \circ f \in R(x)$ on $[a, b]$.

Proof. Choose $\epsilon > 0$. Let $K = \sup \{|\phi(t)| : t \in S\}$, and let $\epsilon' > 0$ be such that $\epsilon'[b - a + 2K] \leqslant \epsilon$. By uniform continuity of ϕ on S, there is $0 < \delta < \epsilon'$ such that $s, t \in S$ and $|s - t| < \delta$ implies $|\phi(s) - \phi(t)| < \epsilon'$. By the integrability of f on $[a, b]$, there is a partition P of $[a, b]$ such that

$$U(P, f) - L(P, f) \leqslant \delta^2.$$

Assume $P = \{x_0, x_1, \ldots, x_n\}$, and let

$$A = \{i : M_i(f) - m_i(f) < \delta\}$$

and

$$B = \{i : M_i(f) - m_i(f) \geqslant \delta\}.$$

Now for $i \in A$ and $s, t \in [x_{i-1}, x_i]$, $|f(s) - f(t)| < \delta$, so

$$|\phi(f(s)) - \phi(f(t))| < \epsilon';$$

that is, $M_i(\phi \circ f) - m_i(\phi \circ f) \leqslant \epsilon'$. For $i \in B$,

$$M_i(f) - m_i(f) \geqslant \delta;$$

hence

$$\begin{aligned}\delta \sum_{i \in B} (x_i - x_{i-1}) &\leqslant \sum_{i \in B} [M_1(f) - m_i(f)](x_i - x_{i-1}) \\ &\leqslant U(P, f) - L(P, f) \leqslant \delta^2.\end{aligned}$$

Thus $\Sigma_{i \in B} (x_i - x_{i-1}) \leqslant \delta$. It now follows that

$$\begin{aligned}&U(P, \phi \circ f) - L(P, \phi \circ f) \\ &\quad = \sum_{i=1}^{n} [M_i(\phi \circ f) - m_i(\phi \circ f)](x_i - x_{i-1}) \\ &\quad = \sum_{i \in A} [M_i(\phi \circ f) - m_i(\phi \circ f)](x_i - x_{i-1}) \\ &\qquad + \sum_{i \in B} [M_i(\phi \circ f) - m_i(\phi \circ f)](x_i - x_{i-1}) \\ &\quad \leqslant \epsilon'(b - a) + 2K\delta \leqslant \epsilon'(b - a + 2K) \leqslant \epsilon.\end{aligned}$$

Thus, by Theorem 5.2, $\phi \circ f \in R(x)$ on $[a, b]$.

It is interesting to note that this last theorem does not follow the pattern we have observed before. It is true that the composition of continuous functions is continuous and the composition of differentiable functions is differentiable; one might conjecture that, if f: $[a, b] \to [c, d]$ and g: $[c, d] \to R$ are such that $f \in R(x)$ on $[a, b]$ and $g \in R(x)$ on $[c, d]$, then $g \circ f \in R(x)$ on $[a, b]$. To see that this is not the case, we shall consider an example.

■ **Example 5.7** Define f: $[0, 1] \to R$ by $f(x) = 0$ if x is irrational and $f(x) = \frac{1}{q}$ if $x = \frac{p}{q}$ with p and q relatively prime nonnegative integers, $q \neq 0$. It has already been shown that $f \in R(x)$ on $[0, 1]$. Define g: $[0, 1] \to R$ by $g(x) = 1$ if $0 < x \leqslant 1$ and $g(0) = 0$. It is left to the reader to show that $g \in R(x)$ on $[a, b]$. Let $h = g \circ f$. Then if x is irrational, $h(x) = 0$, and if x is rational, $h(x) = 1$. As previously observed, h is not integrable on $[0, 1]$. ■

5.12 THEOREM If f: $[a, b] \to R$, g: $[a, b] \to R$, and $f, g \in R(x)$ on $[a, b]$, then $fg \in R(x)$ on $[a, b]$ and $|f| \in R(x)$ on $[a, b]$. Also,

$$\left| \int_a^b f \, dx \right| \leqslant \int_a^b |f| \, dx.$$

Proof. Since f and g are integrable, both functions are bounded, hence there is $M > 0$ such that im f and im g are contained in $[-M, M]$. Define ϕ: $[-2M, 2M] \to R$ by $\phi(x) = x^2$. Clearly ϕ is continuous, and so $\phi \circ (f + g) = (f + g)^2 \in R(x)$ on $[a, b]$ by Theorems 5.11 and 5.9. Similarly, $\phi \circ (f - g) = (f - g)^2 \in R(x)$ on $[a, b]$, hence

$$fg = \frac{1}{4}[(f + g)^2 - (f - g)^2] \in R(x) \text{ on } [a, b].$$

We saw in Chapter 3 that the function $\phi(x) = |x|$ is continuous, hence $\phi \circ f = |f| \in R(x)$ on $[a, b]$. Now $f(x) \leqslant |f|(x)$ and $-f(x) \leqslant |f|(x)$ for all $x \in [a, b]$, so

$$\int_a^b f \, dx \leqslant \int_a^b |f| \, dx \quad \text{and} \quad -\int_a^b f \, dx = \int_a^b - f \, dx \leqslant \int_a^b |f| \, dx.$$

Combining the last two inequalities we obtain $\left| \int_a^b f \, dx \right| \leqslant \int_a^b |f| \, dx.$

Taylor's Theorem has a number of versions; the one we present now gives the remainder as an integral.

5.13 TAYLOR'S THEOREM Let f: $[a, a + h] \to R$ be such that $f^{(n+1)}$ exists and is continuous on $[a, a + h]$. Then, for $x \in [a, a + h]$,

$$f(x) = f(a) + \sum_{k=1}^{n} \frac{f^{(k)}(a)}{k!}(x - a)^k + R_{n+1}(x)$$

where $R_{n+1}(x) = \dfrac{1}{n!}\int_a^x f^{(n+1)}(t)(x - t)^n\,dt.$

Proof. The proof is by induction. For $n = 0$, the theorem is just a restatement of Theorem 5.8 since f' is continuous on $[a, x]$.

Assume the formula holds for $n = m$. All we need to do is to integrate R_{m+1} by parts to obtain R_{m+2} and the next term in the sum. See Exercise 18 for integration by parts. So we have

$$\begin{aligned} R_{m+1} &= \frac{1}{m!}\int_a^x f^{(m+1)}(t)(x - t)^m\,dt \\ &= \frac{1}{m!}\left[f^{(m+1)}(a)\frac{(x - a)^{m+1}}{m + 1} - f^{(m+1)}(x)\frac{(x - x)^{m+1}}{m + 1}\right] \\ &\quad + \frac{1}{(m + 1)!}\int_a^x f^{(m+2)}(t)(x - t)^{m+1}\,dt \\ &= \frac{1}{(m + 1)!}f^{(m+1)}(a)(x - a)^{m+1} + R_{m+2}(x). \end{aligned}$$

5.6 DERIVATIVES OF INTEGRALS

It may be the case that our enthusiasm over Theorems 5.6 and 5.7 forced a premature presentation of the Fundamental Theorem of Integral Calculus. Let us now reconsider that result.

THEOREM If f: $[a, b] \to R$ is differentiable on $[a, b]$ and $f' \in R(x)$ on $[a, b]$, then $\int_a^b f'\,dx = f(b) - f(a)$.

As has already been pointed out, this important theorem reveals the very intimate relation between integration and differentiation. The reader may recall many hours spent in calculating integrals by the use of this theorem. In order to obtain a deeper intuitive feeling for the aspects of these operations, let us take a different point of view. Assume f: $[a, b] \to R$ is differentiable on $[a, b]$. Then a function f': $[a, b] \to R$ exists, and, as noted in various examples, f' need not be continuous; in fact, f' need not even be Riemann-integrable on $[a, b]$. Thus, the operation of differentiation on a differentiable function yields a function that is less nice in the sense that it need not even be continuous. Viewing integration in some fashion as a process inverse to differentiation, one might expect integration to give "nicer" functions. So far, the integral of a function is just a real number, so we must seek some means to use the integral of a function f to create a new function g. That is the purpose of the next two theorems.

Suppose f: $[a, b] \to R$ and $f \in R(x)$ on $[a, b]$. Then, if $a \leqslant c < d \leqslant b$, $f \in R(x)$ on $[c, d]$. If $a \leqslant c \leqslant d \leqslant b$, define

$$\int_d^c f\,dx = -\int_c^d f\,dx \quad \text{and} \quad \int_c^c f\,dx = 0.$$

The following theorem will serve to promote the cause espoused above.

5.14 THEOREM Suppose $f\colon [a, b] \to R$ is bounded and $f \in R(x)$ on $[a, b]$. Define $F(t) = \int_a^t f(x)\,dx$ for $a \leqslant t \leqslant b$. Then

1. F is continuous on $[a, b]$.
2. If f is continuous at x_0, then F is differentiable at x_0 and $F'(x_0) = f(x_0)$.

The function F may be referred to as an indefinite integral of f. Note here that the integrability of f implies the continuity of F and that the continuity of f at x_0 implies the differentiability of F at x_0; hence, F is indeed "nicer" than f.

Proof. Choose $M > 0$ such that $|f(x)| \leqslant M$ for all $x \in [a, b]$. Choose $\epsilon > 0$. Let $\delta = \frac{\epsilon}{M}$. Thus, if $|x - y| < \delta$ and $x, y \in [a, b]$, then

$$\begin{aligned} |F(x) - F(y)| &= \left| \int_a^x f(t)\,dt - \int_a^y f(t)\,dt \right| \\ &= \left| \int_y^x f(t)\,dt \right| \leqslant |x - y| M < \delta M = \epsilon. \end{aligned}$$

Thus, F is continuous on $[a, b]$.

Suppose f is continuous at x_0. Choose $\epsilon > 0$. There is $\delta > 0$ such that $|x_0 - y| < \delta$ with $y \in [a, b]$ implies $|f(x_0) - f(y)| < \frac{\epsilon}{2}$. Thus, if $0 < |x_0 - y| < \delta$,

$$\begin{aligned} \left| \frac{F(x_0) - F(y)}{x_0 - y} - f(x_0) \right| &= \left| \frac{1}{x_0 - y} \int_y^{x_0} f(t)\,dt - f(x_0) \right| \\ &= \frac{1}{|x_0 - y|} \left| \int_y^{x_0} [f(t) - f(x_0)]\,dt \right| \\ &\leqslant \frac{1}{|x_0 - y|} |x_0 - y| \frac{\epsilon}{2} = \frac{\epsilon}{2} < \epsilon. \end{aligned}$$

Therefore, F is differentiable at x_0 and $F'(x_0) = f(x_0)$.

We may use Theorem 5.13 to give an easy proof of a slightly weaker version of the Fundamental Theorem of Integral Calculus.

THEOREM Suppose $f\colon [a, b] \to R$ is differentiable with f' continuous on $[a, b]$. Then $f' \in R(x)$ and

$$\int_a^b f'(x)\,dx = f(b) - f(a).$$

Proof. Define $F: [a, b] \to R$ by $F(x) = \int_a^x f'(t)\,dt$. Since f' is continuous on $[a, b]$, then F is differentiable on $[a, b]$ and $F'(x) = f'(x)$ for all $x \in [a, b]$. This means that there is a constant K such that $F(x) = f(x) + K$ for all $x \in [a, b]$. We evaluate K as follows: $0 = F(a) = f(a) + K$, so $K = -f(a)$. Therefore

$$\int_a^b f'(t) = F(b) = f(b) + K = f(b) - f(a).$$

■ **Example 5.8** Consider the function $f(x) = x^2 \sin \frac{1}{x}$ for $x \neq 0$ and $f(0) = 0$. Now f is differentiable on $[0, 1]$ and $f'(x) = 2x \sin \frac{1}{x} - \cos \frac{1}{x}$ for $x \neq 0$ and $f'(0) = 0$. The function f' is continuous on $(0, 1]$ and bounded; hence, by Exercise 24, f' is integrable on $[0, 1]$. However, the version of the Fundamental Theorem of Integral Calculus given above cannot be used since f' is not continuous on $[0, 1]$. But Theorem 5.13 can be used to determine that

$$\int_0^1 f'(x)\,dx = f(1) - f(0) = \sin 1. \quad ■$$

■ **Example 5.9** For $x > 0$, define $L(x) = \int_1^x \frac{1}{t}\,dt$. Since $f(t) = \frac{1}{t}$ is continuous for $t > 0$, L is differentiable for $t > 0$ and $L'(t) = \frac{1}{t}$. The reader may well remember seeing the *natural logarithm* function defined in this way. See the project at the end of the chapter for a more detailed examination of this function. ■

■ **Example 5.10** For each $x > 0$, define $f(x) = \int_1^{\sqrt{x}} \cos t^2\,dt$. We can consider f as the composition of two functions, $g(x) = \sqrt{x}$ and $h(x) = \int_1^x \cos t^2\,dt$. Then $f = h \circ g$ and we may use the Chain Rule to find f':

$$\begin{aligned} f'(x) &= (h \circ g)'(x) = [h'(g(x))]g'(x) \\ &= [\cos (g(x))^2]g'(x) = (\cos x)\left(\frac{1}{2\sqrt{x}}\right) \quad ■ \end{aligned}$$

As pointed out in the discussion of the Fundamental Theorem of Integral Calculus, the reader's experience with integration has been largely that of being given a function f on $[a, b]$ to find a differentiable function g on $[a, b]$ such that $f = g'$. If f is defined on $[a, b]$ and g is a differentiable function on $[a, b]$ such that $g' = f$, then g is called an *antiderivative* or *primitive* of f. If f is continuous on $[a, b]$, then Theorem 5.4 guarantees that $f \in R(x)$ on $[a, b]$, and then Theorem 5.14 assures us that f has a primitive, namely $g(t) = \int_a^t f\,dx$. It is not true that every integrable function has a primitive, since our results from Chapter 4 show that, if $f = g'$ on $[a, b]$ for some differentiable function g, then f must satisfy certain conditions. For example, the function f, defined by $f(x) = 0$ for $0 \leqslant x \leqslant \frac{1}{2}$ and $f(x) = 1$ for $\frac{1}{2} \leqslant x \leqslant 1$, is Riemann-integrable on $[0, 1]$ but does not have a primitive since the derivative of a differentiable function, by Theorem 4.11, must have the intermediate-value property.

5.7 MEAN-VALUE AND CHANGE-OF-VARIABLES THEOREMS

The next two theorems and the corollary are in a sense analogues of the Mean-Value Theorem for derivatives.

5.15 FIRST MEAN-VALUE THEOREM If $f: [a, b] \to R$ is continuous and $g: [a, b] \to R$ is integrable on $[a, b]$ with $g(x) \geqslant 0$ for all $x \in [a, b]$, then there is $c \in [a, b]$ such that

$$\int_a^b f(x)g(x)\, dx = f(c) \int_a^b g(x)\, dx.$$

Proof. Let $m = \inf \{f(x) : x \in [a, b]\}$ and $M = \sup \{f(x) : x \in [a, b]\}$. Since $g(x) \geqslant 0$ for all $x \in [a, b]$, then $mg(x) \leqslant f(x)g(x) \leqslant Mg(x)$ for all $x \in [a, b]$; hence,

$$m \int_a^b g(x)\, dx \leqslant \int_a^b f(x)g(x)\, dx \leqslant M \int_a^b g(x)\, dx.$$

The only case of interest is that where $\int_a^b g(x)\, dx \neq 0$. In that case,

$$m \leqslant \frac{\int_a^b f(x)g(x)\, dx}{\int_a^b g(x)\, dx} \leqslant M;$$

hence by the continuity of f, there is $c \in [a, b]$ such that

$$f(c) = \frac{\int_a^b f(x)g(x)\, dx}{\int_a^b g(x)\, dx}.$$

If we let $g(x) = 1$ for all $x \in [a, b]$, the following corollary is immediate.

5.16 COROLLARY If $f: [a, b] \to R$ is continuous, then there is $c \in [a, b]$ such that

$$\int_a^b f\, dx = f(c)(b - a).$$

5.17 SECOND MEAN-VALUE THEOREM Suppose $f: [a, b] \to R$ is monotone. Then there is $c \in [a, b]$ such that

$$\int_a^b f\,dx = f(a)(c - a) + f(b)(b - c).$$

Proof. Since f is monotone, $f \in R(x)$ on $[a, b]$. Define $h: [a, b] \to R$ by $h(x) = f(a)(x - a) + f(b)(b - x)$. Now h is continuous and $\int_a^b f\,dx$ is between $h(a) = f(b)(b - a)$ and $h(b) = f(a)(b - a)$, since f is monotone. By the continuity of h there is $c \in [a, b]$ such that $h(c) = \int_a^b f\,dx$. Thus

$$\int_a^b f\,dx = h(c) = f(a)(c - a) + f(b)(b - c).$$

We may use Theorem 5.15 to obtain another version of Taylor's Theorem.

TAYLOR'S THEOREM If $f: [a, a + h] \to R$ is such that $f^{(n+1)}$ exists and is continuous on $[a, a + h]$, then, for each $x \in [a, a + h]$, there is $c \in [a, x]$ such that

$$f(x) = f(a) + \sum_{k=1}^{n} \frac{f^{(k)}(a)}{k!}(x - a)^k + R_{n+1}(x)$$

where $R_{n+1}(x) = \dfrac{f^{(n+1)}(c)}{(n + 1)!}(x - a)^{n+1}$.

Proof. By Theorem 5.13, the formula holds with

$$R_{n+1}(x) = \frac{1}{n!}\int_a^x f^{(n+1)}(t)(x - t)^n\,dt.$$

Theorem 5.15 applies and so there is $c \in [a, x]$ such that

$$R_{n+1}(x) = \frac{1}{n!} f^{(n+1)}(c) \int_a^x (x - t)^n\,dt = \frac{f^{(n+1)}(c)}{(n + 1)!}(x - a)^{n+1}.$$

So far we have included many of the theorems that you encountered when you studied freshman calculus. You were probably given an intuitive idea of how the proofs could be presented and now you have seen the proofs done in a very precise and correct manner. We hope that you have also gained more insight into the Riemann integral. No discussion of integration would be complete without some version of the next theorem. In fact, this theorem was the key to most of the calculations you did with integrals.

The Change-of-Variables Theorem can be stated in a variety of ways. The manner we have chosen is intended to reflect the uses that you have made of the theorem.

5.18 CHANGE-OF-VARIABLES THEOREM Suppose ϕ: $[a, b] \to R$ is differentiable and ϕ' is continuous. Further assume that $\phi([a, b]) = [c, d]$ with $\phi(a) = c$ and $\phi(b) = d$. If f: $[c, d] \to R$ is continuous, then

$$\int_a^b f(\phi(t))\phi'(t)\, dt = \int_c^d f(x)\, dx.$$

Proof. Define $F(u) = \int_c^u f(x)\, dx$ and $G(s) = \int_a^s f(\phi(t))\phi'(t)\, dt$. Since f, ϕ, and ϕ' are continuous, both F and G are differentiable and

$$\begin{aligned} &\text{for } u \in [c, d], \quad F'(u) = f(u), \\ &\text{for } s \in [a, b], \quad G'(s) = f(\phi(s))\phi'(s). \end{aligned}$$

Moreover, $F \circ \phi$ is differentiable on $[a, b]$ and, by the Chain Rule,

$$(F \circ \phi)'(s) = F'(\phi(s))\phi'(s) = f(\phi(s))\phi'(s) = G'(s).$$

This means that there is a constant K such that

$$(F \circ \phi)(s) = G(s) + K$$

for all $s \in [a, b]$. Therefore,

$$0 = F(c) = F(\phi(a)) = G(a) + K = 0 + K.$$

So,

$$\int_c^d f(x)\, dx = F(d) = F(\phi(b)) = G(b) = \int_a^b f(\phi(t))\phi'(t)\, dt.$$

An alternative version of this theorem is given in Exercise 34. A typical application of the Change-of-Variables Theorem is to recognize when the function to be integrated is of the form $(f \circ \phi)(t)\phi'(t)$, or to make changes that yield that form. Follow the next example. (It's probably one you've done before!)

■ **Example 5.11** We will use Theorem 5.18 to evaluate $\int_1^2 (3x^2 - 7)^4(5x)\, dx$. The idea is to recognize this as a function $\phi(x) = 3x^2 - 7$ raised to a power or, in other words, the composition of ϕ with the function that raises a number to the fourth power. Let

$$\phi(x) = 3x^2 - 7 \quad \text{and} \quad f(u) = u^4.$$

Then

$$(f \circ \phi)(x) = (3x^2 - 7)^4, \quad \phi'(x) = 6x, \quad \phi(1) = -4, \quad \text{and} \quad \phi(2) = 5.$$

With a little manipulation, we have

$$\int_1^2 (3x^2 - 7)^4(5x)\, dx = \frac{5}{6}\int_1^2 (3x^2 - 7)^4(6x)\, dx$$

$$= \frac{5}{6}\int_1^2 (f \circ \phi)(x)\phi'(x)\,dx = \frac{5}{6}\int_{-4}^5 f(u)\,du$$

$$= \frac{5}{6}\int_{-4}^5 u^4\,du = \frac{5^5}{6} - \frac{(-4)^5}{6} = \frac{2869}{6}. \qquad ■$$

■ **Example 5.12** Another use of Theorem 5.18 that the reader will recognize is the technique of trigonometric substitution. Follow this example. Consider the function $f(x) = \sqrt{1 - x^2}$ on the interval $[0, 1]$. We wish to compute

$$\int_0^1 \sqrt{1 - x^2}\,dx.$$

The strategy is to apply Theorem 5.18 with $\phi(t) = \sin t$. Now $\phi(0) = 0$ and $\phi(\frac{\pi}{2}) = 1$, ϕ is differentiable and ϕ' is continuous on $[0, \frac{\pi}{2}]$, and f is continuous on $[0, 1]$. Thus, Theorem 5.18 asserts that

$$\int_0^1 \sqrt{1 - x^2}\,dx = \int_0^1 f(x)\,dx = \int_0^{\pi/2} f(\phi(t))\phi'(t)\,dt$$

$$= \int_0^{\pi/2} \sqrt{1 - \sin^2 t}\cos t\,dt = \int_0^{\pi/2} \cos^2 t\,dt$$

$$= \int_0^{\pi/2} \frac{1}{2}(1 + \cos 2t)\,dt = \frac{\pi}{4}.$$

The last step involves some trigonometry, the antiderivative of the cosine function, and the Fundamental Theorem of Integral Calculus. ■

EXERCISES

5.1 THE RIEMANN INTEGRAL

1. Use Theorem 5.2 to prove directly that the function $f(x) = x^3$ is integrable on $[0, 1]$.
2. Use Theorem 5.2 to prove directly that $f(x) = x$ is integrable on $[0, 1]$. Find the integral of f by finding a number A such that $L(P, f) \leq A \leq U(P, f)$ for all partitions of $[0, 1]$.
3. Define $f(x) = x$ if x is rational and $f(x) = 0$ if x is irrational. Compute $\overline{\int_0^1} f\,dx$ and $\underline{\int_0^1} f\,dx$. Is f integrable on $[0, 1]$? You may wish to look at the results of Exercise 2.
4. A set $A \subset [0, 1]$ is dense in $[0, 1]$ iff every open interval that intersects $[0, 1]$ contains a point of A. Suppose f: $[0, 1] \to R$ is integrable and $f(x) = 0$ for all $x \in A$ with A dense in $[0, 1]$. Show that $\int_0^1 f(x)\,dx = 0$.
5. Define f: $[0, 2] \to R$ by $f(x) = 1$ for $0 \leq x \leq 1$ and $f(x) = 2$ for $1 < x \leq 2$. Show that $f \in R(x)$ on $[0, 2]$ and compute the integral.

5.2 CLASSES OF INTEGRABLE FUNCTIONS

6. If $f: [a, b] \to R$ is decreasing, prove $f \in R(x)$ on $[a, b]$.

***7.** Suppose $g: [a, b] \to R$ is bounded and continuous except at $x_0 \in (a, b)$. Prove that $g \in R(x)$ on $[a, b]$. See Exercises 24 and 25 for generalizations of this result.

8. Find the integral of $f(x) = x$ on $[1, 3]$ using the techniques of Example 5.4.

9. Assume $f: [a, b] \to R$ is continuous and $f(x) \geq 0$ for all $x \in [a, b]$. Prove that if $\int_a^b f\,dx = 0$, then $f(x) = 0$ for all $x \in [a, b]$.

5.3 RIEMANN SUMS

***10.** Prove Theorem 5.7.

11. Show that, for $a > 1$ and $b > 1$, the function $f(x) = \frac{1}{x}$ is integrable on $[1, a]$ and on $[b, ab]$. Use the results of Section 5.3 to show that $\int_1^a \frac{1}{x}\,dx = \int_b^{ab} \frac{1}{x}\,dx$.

12. Suppose $f \in R(x)$ on $[0, 1]$. Define $a_n = \frac{1}{n}\sum_{k=1}^{n} f(\frac{k}{n})$ for all n. Prove $\{a_n\}_{n=1}^{\infty}$ converges to $\int_0^1 f\,dt$.

13. Suppose $f: [a, b] \to R$ is bounded and for each $\epsilon > 0$ there is a partition P such that for any refinements Q_1 and Q_2 of P, regardless of how marked, $|S(Q_1, f) - S(Q_2, f)| < \epsilon$. Prove that f is integrable on $[a, b]$. (Note that this applies to different markings of the same partition.)

14. Suppose f is integrable on $[-b, b]$ and f is an odd function—that is, $f(-t) = -f(t)$ for all $t \in [-b, b]$. Prove that $\int_{-b}^{b} f\,dx = 0$. If f is even—that is, $f(-t) = f(t)$ for all $t \in [-b, b]$—prove that

$$\int_{-b}^{b} f\,dx = 2\int_0^b f\,dx.$$

15. Suppose that $f: R \to R$ is periodic and integrable on every closed interval. If p is the period of f, prove that for any $a \in R$,

$$\int_0^p f\,dx = \int_a^{a+p} f\,dx.$$

5.4 THE FUNDAMENTAL THOEREM OF INTEGRAL CALCULUS

16. Use the Fundamental Theorem of Integral Calculus to compute the following:

a. $\int_0^3 (x^2 - x)\,dx$ b. $\int_{-2}^4 (1 - x^3 - x^2)\,dx$ c. $\int_0^{\pi/2} x \sin x^2\,dx$

17. Define $f: [0, 2] \to R$ by $f(x) = 2x - x^2$ for $0 \leq x \leq 1$ and $f(x) = (x - 2)^2$ for $1 < x \leq 2$. Prove that f is integrable on $[0, 2]$ and find the integral of f over $[0, 2]$. Do not use Theorem 5.10, but rather find the integral by methods similar to those used in the proof of Theorem 5.8.

5.5 ALGEBRA OF INTEGRABLE FUNCTIONS

***18.** Suppose that f and g are differentiable on $[a, b]$ and f' and g' are integrable on $[a, b]$. Prove that $f'g$ and $g'f$ are integrable on $[a, b]$ and that

$$\int_a^b f'g\,dx = f(b)g(b) - f(a)g(a) - \int_a^b g'f\,dx.$$

Of course, this is the *integration-by-parts formula.*

19. Suppose that $f \in R(x)$ on $[a, b]$ and $\frac{1}{f}$ is bounded on $[a, b]$. Prove that $\frac{1}{f} \in R(x)$ on $[a, b]$.

20. Suppose that $f \in R(x)$ on $[a, b]$ and $f(x) \geqslant 0$ for all $x \in [a, b]$. Prove that $\sqrt{f} \in R(x)$ on $[a, b]$.

21. Use Exercise 18 to calculate $\int_0^{\pi/4} x \cos x\,dx$. You may assume what you need concerning the derivatives of the trigonometric functions.

22. Suppose f is continuous on $[0, 1]$. Define $g_n(x) = f(x^n)$ for $n = 1, 2, \ldots$. Prove that $\left\{\int_0^1 g_n(x)\,dx\right\}_{n=1}^{\infty}$ converges to $f(0)$.

23. Define $\gamma_n = 1 + \frac{1}{2} + \frac{1}{3} + \cdots + \frac{1}{n} - \int_1^n \frac{1}{t}\,dt$. Prove that $\{\gamma_n\}_{n=1}^{\infty}$ converges.

***24.** Suppose $g\colon [a, b] \to R$ is bounded and continuous except at $x_1, \ldots, x_n \in [a, b]$. Prove that $g \in R(x)$ on $[a, b]$.

***25.** Suppose $\{a_n\}_{n=1}^{\infty}$ is a sequence of members of $[a, b]$ converging to x_0 in $[a, b]$. If f is bounded on $[a, b]$ and continuous on $[a, b]$ except at x_0 and the points of the sequence $\{a_n\}_{n=1}^{\infty}$, prove that f is integrable on $[a, b]$.

26. Prove that $\frac{1}{3\sqrt{2}} \leqslant \int_0^1 \frac{x^2}{\sqrt{1 + x^2}}\,dx \leqslant \frac{1}{3}$. [*Hint:* Theorem 5.9 (b) and carefully chosen functions should do the trick.]

27. Suppose f and g are integrable on $[a, b]$. Define $h(x) = \max\{f(x), g(x)\}$. Prove that h is integrable on $[a, b]$.

5.6 DERIVATIVES OF INTEGRALS

28. Suppose $f\colon [0, 1] \to R$ is continuous and $\int_0^x f(t)\,dt = \int_x^1 f(t)\,dt$ for all $x \in [0, 1]$. Prove that $f(x) = 0$ for all $x \in [0, 1]$.

29. Suppose that f and g are continuous on $[a, b]$ and $\int_a^b f(x)\,dx = \int_a^b g(x)\,dx$. Prove that there is $c \in [a, b]$ such that $f(c) = g(c)$.

30. Find f' where f is defined on $[0, 1]$ as indicated:

a. $f(x) = \int_0^x \sqrt{t^2 + 1}\,dt$

b. $f(x) = \int_x^1 \cos \frac{1}{t + 1}\,dt$

c. $f(x) = \int_{x^2}^{2x} \sin t^2\,dt$

d. $f(x) = \int_x^{\sqrt{x}} \frac{1}{1 + t^3}\,dt$

31. Let $f\colon R \to R$ be continuous and $\delta > 0$. Define $g(t) = \int_{t-\delta}^{t+\delta} f(x)\,dx$ for all $t \in R$. Prove that g is differentiable and compute g'.

***32.** Suppose $f\colon [a, b] \to R$ is continuous and $g\colon [c, d] \to [a, b]$ is differentiable. Define $F(x) = \int_a^{g(x)} f(t)\,dt$ for $x \in [c, d]$. Prove that F is differentiable and compute F'.

5.7 MEAN-VALUE AND CHANGE-OF-VARIABLE THEOREMS

33. Suppose $f: R \to R$ is continuous and has period p, so that $f(x + p) = f(x)$ for all $x \in R$. Show that $\int_x^{x+p} f(t)\, dt$ is independent of x in that, for all x, y,

$$\int_x^{x+p} f(t)\, dt = \int_y^{y+p} f(t)\, dt.$$

Show, then, that $\int_0^p [f(x + a) - f(x)]\, dx = 0$ for any real number a. Conclude that for any real number a, there is x such that $f(x + a) = f(x)$.

***34.** Prove the following variation on Theorem 5.18. Assume that $\phi: [a, b] \to R$ is differentiable, 1–1, and increasing with $\phi(a) = c$ and $\phi(b) = d$. If $f: [c, d] \to R$ is integrable on $[c, d]$ and $(f \circ \phi)\phi'$ is also integrable on $[a, b]$, then

$$\int_c^d f(x)\, dx = \int_a^b f(\phi(t))\phi'(t)\, dt.$$

35. Use Theorem 5.18 to evaluate the integrals:

a. $\int_0^3 \sqrt[3]{1 + x^2}\, x\, dx$

b. $\int_1^4 \frac{(\sqrt{x} + 2)^3}{\sqrt{x}}\, dx$

c. $\int_1^{\sqrt{3}} \frac{\sqrt{x^2 - 9}}{x}\, dx$

d. $\int_0^1 \frac{x^2}{\sqrt{1 - x^2}}\, dx$

[You may want to consult your calculus book before trying (c) and (d).]

36. Use Theorem 5.18 to establish the result in Exercise 11.

37. Use Theorem 5.18 to establish the result proven in Example 5.5.

MISCELLANEOUS

38. Assume $f: [a, b] \to R$ is continuous, $f(x) \geq 0$ for all $x \in [a, b]$, and $M = \sup \{f(x) : x \in [a, b]\}$. Show that $\left\{\left[\int_a^b [f(x)]^n\, dx\right]^{1/n}\right\}_{n=1}^{\infty}$ converges to M.

39. For each positive integer n, define $a_n = \frac{1}{n}\left[\left(\frac{1}{n}\right)^2 + \left(\frac{2}{n}\right)^2 + \cdots + \left(\frac{n}{n}\right)^2\right]$. Find the limit of the sequence $\{a_n\}_{n=1}^{\infty}$.

40. For each positive integer n, define $a_n = \frac{1}{n}\left[\sin\frac{\pi}{n} + \sin\frac{2\pi}{n} + \cdots + \sin\frac{n\pi}{n}\right]$. Find the limit of the sequence $\{a_n\}_{n=1}^{\infty}$.

41. If f and g are integrable on $[a, b]$, show that

$$\left[\int_a^b fg\, dx\right]^2 \leq \int_a^b f^2\, dx \int_a^b g^2\, dx.$$

[*Hint:* For all α and β, $\int_a^b (\alpha f + \beta g)^2\, dx \geq 0$. Let $\alpha = \int_a^b g^2\, dx$ and $\beta = -\int_a^b fg\, dx$.]

42. Use Exercise 41 to prove the Minkowski inequality—that is, if f and g are integrable on $[a, b]$, then

$$\left[\int_a^b [f(x) + g(x)]^2\, dx\right]^{1/2} \leq \left[\int_a^b [f(x)]^2\, dx\right]^{1/2} + \left[\int_a^b [g(x)]^2\, dx\right]^{1/2}.$$

PROJECT

The purpose of this project is to use the integral to define the natural logarithm function, $\log x$, and to develop some of the properties of that function.

Define $\log x = \displaystyle\int_1^x \frac{1}{t}\, dt$ for $0 < x$. For convenience, we will write $\log x = L(x)$.

1. Prove that L is continuous, differentiable, and 1–1.
2. Prove that L is increasing.
3. Prove that $L(ab) = L(a) + L(b)$ for all $a, b > 0$.
4. Prove that $L(x^n) = nL(x)$ for all $x > 0$, $n \in Z$.

Comment: One possible approach to (3) is to define $f(x) = L(ax)$ and show that $f'(x) = L'(x)$, hence $f(x) = L(ax) = L(x) + k$ and then show that $k = L(a)$. A similar approach will supply a proof for (4).

5. Prove that $L\left(\dfrac{a}{b}\right) = L(a) - L(b)$ for $a, b > 0$.
6. Prove that im $L = R$. Since $L\left(\dfrac{1}{x}\right) = -L(x)$ for $1 < x$, and $L(1) = 0$, it suffices to prove that for any $y > 0$, there is $x > 1$ such that $L(x) = y$. Since L is continuous, you need to show that there is $n \in J$ such that $L(n) > y$ and then use the Intermediate-Value Theorem. It might be helpful for you to look ahead to Example 6.3.

We know that $\left\{\left(1 + \dfrac{1}{n}\right)^n\right\}_{n=1}^{\infty}$ converges and we choose to call the limit e. Now we will show that $L(e) = 1$. It suffices to show that $\left\{L\left(\left(1 + \dfrac{1}{n}\right)^n\right)\right\}_{n=1}^{\infty}$ converges to 1. Do you see why?

7. Prove that for each $n \in J$, $\dfrac{n}{n+1} \leq nL\left(1 + \dfrac{1}{n}\right) \leq 1$. You will want to write $L\left(1 + \dfrac{1}{n}\right)$ as an integral and use Theorem 5.9 (c).
8. Finish the proof that $L(e) = 1$.

9. Let E be the inverse of L (remember, L is 1–1). Since im $L = R$, then dom $E = R$ and im $E = \{x : x \in R, x > 0\}$. Prove that E is continuous, differentiable, 1–1, and increasing. Find E'. Be careful; this is a bit tricky. You cannot use Theorem 3.12 directly since $\{x : x \in R, x > 0\}$ is not compact. However, you can modify the proof of 3.12 to meet your needs. Similar observations are appropriate concerning Theorem 4.14. On the other hand, you might find some assistance in the miscellaneous exercises in Chapters 2, 3, and 4.
10. Prove that E behaves like an exponential function—that is, $E(x + y) = E(x)E(y)$ for all $x, y \in R$.
11. We have some understanding of x^r where x is any real number greater than zero and r is any rational number. Now we can define x^y for any real number $x > 0$ and any real number y. Define $x^y = E(yL(x))$ or, in more familiar notation, $x^y = e^{y \log x}$. Show that $f(x) = x^x$ is differentiable and find f' (x is assumed to be a positive real number).

CHAPTER 6

Infinite Series

Most students of mathematics who have completed calculus feel competent in the techniques of differentiation and integration, but they may panic at the mention of infinite series. However, the notions of convergence and divergence of infinite series are actually quite easy to assimilate if presented properly. Since the question of convergence of an infinite series depends on the convergence of a certain sequence, a brief review of sequences, in Chapter 1, may help you to prepare for what is to come.

Finite sums of real numbers are familiar objects; our purpose here is to determine when we should assign a real number to a string of symbols such as $a_1 + a_2 + a_3 + \cdots$. The approach is the usual one—that of looking at the partial sums of this "infinite sum."

6.1 CONVERGENCE OF INFINITE SERIES

DEFINITION An *infinite series* is a pair $(\{a_n\}_{n=1}^{\infty}, \{S_n\}_{n=1}^{\infty})$ where $\{a_n\}_{n=1}^{\infty}$ is a sequence of real numbers and $S_n = \sum_{k=1}^{n} a_k$ for $n = 1, 2, \ldots$. The number a_n is called the *nth term* of the series, and S_n is called the *nth partial sum* of the series.

In keeping with the more customary notation, we may denote the infinite series $(\{a_n\}_{n=1}^{\infty}, \{S_n\}_{n=1}^{\infty})$ by $\sum_{n=1}^{\infty} a_n$ or by $a_1 + a_2 + \cdots + a_n + \cdots$. It is often convenient to index the terms of an infinite series beginning with an integer other than 1. As our discussion unfolds, it will be clear that questions of convergence are independent of whether we index the terms beginning with $n = 1$ or $n = p$ for some other integer p.

DEFINITION The infinite series $(\{a_n\}_{n=1}^{\infty}, \{S_n\}_{n=1}^{\infty})$ *converges* if the sequence $\{S_n\}_{n=1}^{\infty}$ converges. If $\{S_n\}_{n=1}^{\infty}$ converges to S, we write

$$\sum_{n=1}^{\infty} a_n = S.$$

The number S is called the *sum* of the infinite series $\Sigma_{n=1}^{\infty} a_n$. If the sequence $\{S_n\}_{n=1}^{\infty}$ does not converge, we say that the infinite series $\Sigma_{n=1}^{\infty} a_n$ *diverges*.

Perhaps a word of apology should be offered for using $\Sigma_{n=1}^{\infty} a_n$ as a name both for an infinite series and also for the real number that is the limit of the sequence of partial sums when the series converges. However, this abuse conforms to convention and the reader's experiences. Note that, if the series does not converge, we *do not* use $\Sigma_{n=1}^{\infty} a_n$ to denote a real number.

Let us first consider an infinite series $\Sigma_{n=1}^{\infty} a_n$ where $a_n = 0$ for $n \geqslant N + 1$. Then, for $n \geqslant N$, $S_n = S_N$; hence $\{S_n\}_{n=1}^{\infty}$ converges to $S_N = a_1 + a_2 + \cdots + a_N$. In this case, the infinite series is in reality a finite sum, a fact that should certainly be no surprise to the reader. Suppose now that $a_n > 0$ for all n; then $\{S_n\}_{n=1}^{\infty}$ is increasing. Hence, in order for $\Sigma_{n=1}^{\infty} a_n$ to converge, it is necessary that $\{S_n\}_{n=1}^{\infty}$ be bounded. Now $S_{n+1} = S_n + a_{n+1}$ for each n; so for $\{S_n\}_{n=1}^{\infty}$ to be bounded, it is necessary that the terms a_n get small "fast enough" as n gets large. To illustrate what we mean by "fast enough," we shall consider several examples.

■ **Example 6.1** Consider the infinite series $\Sigma_{n=1}^{\infty} \log\left(\frac{n+1}{n}\right)$. (We shall assume the usual properties of the function $\log x$.) Now

$$S_n = \log 2 + \log\frac{3}{2} + \log\frac{4}{3} + \cdots + \log\frac{n+1}{n} = \log(n+1).$$

The sequence $\{S_n\}_{n=1}^{\infty}$ is unbounded and, hence, not convergent. Note that $\left\{\frac{n+1}{n}\right\}_{n=1}^{\infty}$ converges to 1, so by the continuity of the log function,

$$\left\{\log\frac{n+1}{n}\right\}_{n=1}^{\infty}$$

converges to $\log 1 = 0$. In this case, the nth term of the series gets small as n gets large, but not "fast enough." ■

■ **Example 6.2** Consider now the series $\Sigma_{n=1}^{\infty} \frac{1}{n(n+1)}$. For each n,

$$S_n = \frac{1}{2} + \frac{1}{2\cdot 3} + \frac{1}{3\cdot 4} + \cdots + \frac{1}{n(n+1)} = 1 - \frac{1}{n+1}$$

since for all k,

$$\frac{1}{k(k+1)} = \frac{1}{k} - \frac{1}{k+1}.$$

Clearly $\{S_n\}_{n=1}^{\infty}$ converges to 1, and we write

$$\sum_{n=1}^{\infty} \frac{1}{n(n+1)} = 1. \quad ■$$

In Example 6.2, the nth term of the series converged to zero "fast enough." Unfortunately, there are no simple criteria for determining whether or not a series converges merely by examining the nth term of the series. In this chapter, we present some standard methods for determining the convergence of infinite series.

Suppose now that $\Sigma_{n=1}^{\infty} a_n$ is convergent, which means that $\{S_n\}_{n=1}^{\infty}$ is convergent. By our previous discussion of sequences, $\{S_n\}_{n=1}^{\infty}$ is convergent iff it is Cauchy.

6.1 THEOREM The infinite series $\Sigma_{n=1}^{\infty} a_n$ converges iff for each $\epsilon > 0$ there is N such that if $n \geqslant N$ and $p \geqslant 0$, then

$$|a_n + a_{n+1} + \cdots + a_{n+p}| < \epsilon.$$

Proof. Suppose $\Sigma_{n=1}^{\infty} a_n$ converges. Then the sequence $\{S_n\}_{n=1}^{\infty}$ converges and, hence, is Cauchy. Choose $\epsilon > 0$. There is N such that $m, n \geqslant N$ implies that $|S_m - S_n| < \epsilon$. Now if $n \geqslant N + 1$ and $p \geqslant 0$, then $n - 1 \geqslant N$ and $n + p \geqslant N$; hence,

$$|a_n + a_{n+1} + \cdots + a_{n+p}| = |S_{n+p} - S_{n-1}| < \epsilon.$$

Assume the condition holds, and choose $\epsilon > 0$. There is N such that for $n \geqslant N$ and $p \geqslant 0$, $|a_n + a_{n+1} + \cdots + a_{n+p}| < \epsilon$. Choose $r \geqslant n \geqslant N$. Then

$$|S_r - S_n| = |a_{n+1} + \cdots + a_r| < \epsilon.$$

Hence, $\{S_n\}_{n=1}^{\infty}$ is Cauchy and the series converges.

Let us pause to ponder the content of Theorem 6.1. First of all, the convergence or divergence of an infinite series is independent of whether we index the terms beginning with $n = 1$ or with $n = p$ for some other integer p. Secondly, the convergence is unaffected by changing finitely many terms of the series; although if the series converges, the limit of the partial sums may be altered. Now if $\Sigma_{n=1}^{\infty} a_n$ converges and k is a positive integer, then $\Sigma_{n=k}^{\infty} a_n$ converges; in fact, given $\epsilon > 0$, there is N such that

$$|a_N + a_{N+1} + \cdots + a_{N+p}| < \epsilon$$

for all $p \geqslant 0$, but

$$a_N + a_{N+1} + \cdots + a_m = T_m$$

is the mth partial sum for the convergent infinite series $\Sigma_{n=N}^{\infty} a_n$; hence, $|\Sigma_{n=N}^{\infty} a_n| \leqslant \epsilon$.

6.2 COROLLARY If $\Sigma_{n=1}^{\infty} a_n$ converges, then $\{a_n\}_{n=1}^{\infty}$ converges to zero.

Proof. Choose $\epsilon > 0$. By Theorem 6.1, there is N such that, for $n \geqslant N$ and $p \geqslant 0$, we have $|a_n + a_{n+1} + \cdots + a_{n+p}| < \epsilon$. In particular, for $p = 0$, $|a_n| < \epsilon$. Hence, $\{a_n\}_{n=1}^{\infty}$ converges to zero.

Corollary 6.2 is presented with mixed emotions. Its result is very useful, since it allows one to observe quite easily that certain series diverge—namely, those series $\Sigma_{n=1}^{\infty} a_n$ where $\{a_n\}_{n=1}^{\infty}$ does not converge to zero. Unfortunately, the result is frequently misused, for many students assume that the converse is true—that if $\{a_n\}_{n=1}^{\infty}$ converges to zero, then $\Sigma_{n=1}^{\infty} a_n$ converges. This converse is definitely false, as we have already seen by considering

$$\sum_{n=1}^{\infty} \log\left(\frac{n+1}{n}\right).$$

Here

$$\left\{\log\left(\frac{n+1}{n}\right)\right\}_{n=1}^{\infty}$$

converges to zero, but the series

$$\sum_{n=1}^{\infty} \log\left(\frac{n+1}{n}\right)$$

diverges. The following example emphasizes this fact and will be of use later.

■ **Example 6.3** Consider the infinite series $\Sigma_{n=1}^{\infty} \frac{1}{n}$, often referred to as the *harmonic series*. Since $\frac{1}{n} > 0$ for each n, the sequence $\{S_n\}_{n=1}^{\infty}$ is increasing, and we shall show it is unbounded. If $n = 2^k$, then

$$\begin{aligned} S_n &= 1 + \frac{1}{2} + \frac{1}{3} + \cdots + \frac{1}{2^k} \\ &= 1 + \frac{1}{2} + \left(\frac{1}{3} + \frac{1}{4}\right) + \left(\frac{1}{5} + \frac{1}{6} + \frac{1}{7} + \frac{1}{8}\right) + \cdots + \left(\frac{1}{2^{k-1}+1} + \cdots + \frac{1}{2^k}\right) \\ &\geqslant 1 + \frac{1}{2} + 2\left(\frac{1}{4}\right) + 4\left(\frac{1}{8}\right) + \cdots + 2^{k-1}\left(\frac{1}{2^k}\right) = 1 + \frac{k}{2}. \end{aligned}$$

Thus, $\{S_n\}_{n=1}^{\infty}$ is unbounded, so $\Sigma_{n=1}^{\infty} \frac{1}{n}$ is divergent. ■

In the study of infinite series, it is useful to have a great many examples at hand. Such examples help to develop the intuition and, as will be seen later, may sometimes be used in testing infinite series for convergence. Before continuing further, we shall present one more classic example.

■ **Example 6.4** The series $\Sigma_{n=0}^{\infty} r^n$ is called the *geometric series* with *ratio r.* For $n \geqslant 0$,

$$S_n = 1 + r + r^2 + \cdots + r^n = \frac{1 - r^{n+1}}{1 - r}$$

for all $r \neq 1$. Now $\{r^{n+1}\}_{n=1}^{\infty}$ converges iff $-1 < r \leqslant 1$; and if $|r| < 1$, $\{r^{n+1}\}_{n=1}^{\infty}$ converges to 0. For $r = 1$, $S_n = 1 + 1 + \cdots + 1 = n + 1$; hence, $\{S_n\}_{n=1}^{\infty}$ is not convergent. To summarize, $\Sigma_{n=0}^{\infty} r^n$ converges to $\dfrac{1}{1-r}$ for $|r| < 1$ and diverges for $|r| \geqslant 1$. ■

■ **Example 6.5** This example illustrates the fact that we have had experience with infinite series before encountering the idea in calculus. Of course, we are referring to *repeating decimals.* Some rational numbers have a terminating decimal representation, for example, $\frac{1}{4} = 0.25$. But for some rational numbers there is no terminating decimal representation, and we commonly write and refer to them as repeating decimals. For example, we would write $\frac{2}{11} = 0.1818\overline{18}$ with the bar over the digits 18 indicating that those digits repeat forever! More compactly, we would write $\frac{2}{11} = 0.\overline{18}$ which implies that

$$\frac{2}{11} = \frac{18}{100} + \frac{18}{10000} + \frac{18}{1000000} + \cdots + \frac{18}{(100)^n} + \cdots$$

This means that $\frac{2}{11}$ is the sum of the infinite series $\Sigma_{n=1}^{\infty} \dfrac{18}{100^n}$. For each n, the nth partial sum is

$$\frac{18}{100}\left[1 + \frac{1}{100} + \cdots + \left(\frac{1}{100}\right)^n\right] = \frac{18}{100}\left[\frac{1 - \left(\dfrac{1}{100}\right)^{n+1}}{1 - \dfrac{1}{100}}\right].$$

Since $0 \leqslant \dfrac{1}{100} < 1$, the sequence of partial sums converges to $\dfrac{18}{100}\left(\dfrac{1}{1 - \dfrac{1}{100}}\right)$ $= \dfrac{18}{100}\left(\dfrac{100}{99}\right) = \dfrac{2}{11}$ and our faith in repeating decimals is reinforced. Notice that the series for the repeating decimal is just a multiple of a geometric series.

■

In Example 6.5, we were dealing with a multiple of a series already examined in detail, the geometric series. It would have been helpful to have had the following theorem at our disposal.

6.3 THEOREM Suppose $\Sigma_{n=1}^{\infty} a_n$ and $\Sigma_{n=1}^{\infty} b_n$ converge and that α and β are real numbers. Then $\Sigma_{n=1}^{\infty} (\alpha a_n + \beta b_n)$ converges and

$$\sum_{n=1}^{\infty} (\alpha a_n + \beta b_n) = \alpha \sum_{n=1}^{\infty} a_n + \beta \sum_{n=1}^{\infty} b_n.$$

Proof. The proof is left as Exercise 10.

■ **Example 6.6** Consider the series $\Sigma_{n=1}^{\infty}\left(\frac{3}{4^n}+\frac{5}{n(n+1)}\right)$. Now the series $\Sigma_{n=0}^{\infty}\frac{1}{4^n}$ converges to $\frac{4}{3}$ (see Example 6.4) and $\Sigma_{n=1}^{\infty}\frac{1}{n(n+1)}$ converges to 1 (see Example 6.2), so $\Sigma_{n=1}^{\infty}\left(\frac{3}{4^n}+\frac{5}{n(n+1)}\right)$ which is $\Sigma_{n=1}^{\infty}\left(\frac{3}{4}\cdot\frac{1}{4^{n-1}}+5\cdot\frac{1}{n(n+1)}\right)$ converges to $\frac{3}{4}\left(\frac{4}{3}\right)+5(1)=6$. Notice that we used the fact that $\Sigma_{n=0}^{\infty}\frac{1}{4^n}$ and $\Sigma_{n=1}^{\infty}\frac{1}{4^{n-1}}$ are the same series. ■

6.2 ABSOLUTE CONVERGENCE AND THE COMPARISON TEST

So far, in most of the examples we have considered, the terms of the infinite series were nonnegative. Let us consider now the infinite series

$$\sum_{n=1}^{\infty}(-1)^n\frac{1}{n}.$$

Shortly it will be shown that this series converges; in fact, this will follow from the alternating series test, which the reader may recall from elementary calculus. It has been remarked earlier that if $a_n \geqslant 0$ for all n, the infinite series $\Sigma_{n=1}^{\infty} a_n$ converges if the nth term tends to zero "fast enough." Now $\Sigma_{n=1}^{\infty}\frac{1}{n}$ is divergent; hence $\left\{\frac{1}{n}\right\}_{n=1}^{\infty}$ does not converge to zero "fast enough." However, $\Sigma_{n=1}^{\infty}(-1)^n\frac{1}{n}$ does converge, so it is clear that the convergence depends somehow on the fact that the terms alternate in sign. We now give a definition that we shall use to investigate this behavior further.

DEFINITION An infinite series $\Sigma_{n=1}^{\infty} a_n$ *converges absolutely* iff $\Sigma_{n=1}^{\infty}|a_n|$ converges. If $\Sigma_{n=1}^{\infty} a_n$ converges but $\Sigma_{n=1}^{\infty}|a_n|$ diverges, then $\Sigma_{n=1}^{\infty} a_n$ is said to *converge conditionally.*

Thus, $\Sigma_{n=1}^{\infty}(-1)^n\frac{1}{n}$ converges conditionally. The series

$$\sum_{n=1}^{\infty}(-1)^n\frac{1}{n(n+1)}$$

converges absolutely since we have already demonstrated that

$$\sum_{n=1}^{\infty}\frac{1}{n(n+1)}$$

converges. It is automatic that every absolutely convergent infinite series is convergent.

6.4 THEOREM Suppose $\Sigma_{n=1}^{\infty} a_n$ is absolutely convergent. Then $\Sigma_{n=1}^{\infty} a_n$ is convergent.

Proof. Choose $\epsilon > 0$. Now since $\Sigma_{n=1}^{\infty} a_n$ is absolutely convergent, $\Sigma_{n=1}^{\infty} |a_n|$ converges. Thus, there is N such that $n \geqslant N$ and $p \geqslant 0$ implies that

$$||a_n| + \cdots + |a_{n+p}|| < \epsilon.$$

But $|a_n + a_{n+1} + \cdots + a_{n+p}| \leqslant |a_n| + \cdots + |a_{n+p}| < \epsilon$; hence, by Theorem 6.1, $\Sigma_{n=1}^{\infty} a_n$ converges.

Some of the distinctions between absolute convergence and conditional convergence shall be brought out as our tale unfolds.

Keep in mind that an infinite series $\Sigma_{n=1}^{\infty} a_n$ of nonnegative terms converges if the terms get small "fast enough." Thus, the infinite series $\Sigma_{n=1}^{\infty} a_n$ converges absolutely if the sequence $\{|a_n|\}_{n=1}^{\infty}$ converges to zero "fast enough." This notion of "fast enough" is quite vague, and we shall try to make this notion more meaningful. To this end, consider two infinite series $\Sigma_{n=1}^{\infty} a_n$ and $\Sigma_{n=1}^{\infty} b_n$, where $a_n \geqslant 0$ and $b_n \geqslant 0$ for all n. If $\Sigma_{n=1}^{\infty} b_n$ converges, then the sequence $\{b_n\}_{n=1}^{\infty}$ converges to zero "fast enough." Suppose now that there is N such that for $n \geqslant N$, $0 \leqslant a_n \leqslant b_n$. Then it seems plausible to conjecture that $\{a_n\}_{n=1}^{\infty}$ converges to zero as fast as $\{b_n\}_{n=1}^{\infty}$, which is "fast enough." These remarks yield the basis for Theorem 6.5, the "comparison test." In other words, we shall try to determine convergence or divergence of a series by comparing it to one whose behavior is already known. Since the comparison must take place between infinite series of nonnegative numbers, the comparison test will yield information only concerning absolute convergence. The success of this test depends on the user's stock of examples. Therefore, it is important to begin to build this stock. The observant reader will note that the comparison test is the basis for the limit-comparison, ratio, and root tests to be presented later. In the latter two cases, the series to be examined are compared to the geometric series $\Sigma_{n=0}^{\infty} r^n$.

6.5 THEOREM (Comparison Test) Suppose $\Sigma_{n=1}^{\infty} a_n$ and $\Sigma_{n=1}^{\infty} b_n$ are infinite series with $b_n \geqslant 0$ for all n. Then

1. If $\Sigma_{n=1}^{\infty} b_n$ converges and there is N_0 such that $n \geqslant N_0$ implies $|a_n| \leqslant b_n$, then $\Sigma_{n=1}^{\infty} a_n$ converges absolutely.
2. If $\Sigma_{n=1}^{\infty} b_n$ diverges and there is N_0 such that $n \geqslant N_0$ implies $b_n \leqslant |a_n|$, then $\Sigma_{n=1}^{\infty} |a_n|$ diverges.

(See Exercise 16 for a generalization of this test.)

Proof. Suppose the hypotheses of (1) hold. Thus, $\Sigma_{n=1}^{\infty} b_n$ converges and there is N_0 such that $n \geqslant N_0$ implies $|a_n| \leqslant b_n$. Since $\Sigma_{n=1}^{\infty} b_n$ converges, given $\epsilon > 0$, there is N such that $n \geqslant N$ and $p \geqslant 0$ imply that $b_n + b_{n+1} + \cdots + b_{n+p} < \epsilon$. Let

$$N_1 = \max \{N, N_0\}.$$

Then if $n \geqslant N_1$ and $p \geqslant 0$,

$$|a_n| + \cdots + |a_{n+p}| \leqslant b_n + b_{n+1} + \cdots + b_{n+p} < \epsilon.$$

Thus, $\Sigma_{n=1}^{\infty} |a_n|$ converges, and $\Sigma_{n=1}^{\infty} a_n$ converges absolutely.

The proof of (2) can be obtained most easily by using the result just obtained. If $\Sigma_{n=1}^{\infty} |a_n|$ converges and there is N_0 such that $n \geqslant N_0$ implies $|b_n| = b_n \leqslant |a_n|$, then since $|a_n| \geqslant 0$, the result of (1) guarantees that $\Sigma_{n=1}^{\infty} |b_n|$ converges, contrary to assumption. Thus, (2) holds.

Let us give several examples to illustrate Theorem 6.5.

■ **Example 6.7** Consider the series

$$\sum_{n=1}^{\infty} \frac{1}{(n+1)^2}.$$

We shall choose to compare this series with the series

$$\sum_{n=1}^{\infty} \frac{1}{n(n+1)},$$

because this is one of the few examples we know much about and because for all n,

$$\frac{1}{(n+1)^2} \leqslant \frac{1}{n(n+1)}.$$

Hence, by Theorem 6.5,

$$\sum_{n=1}^{\infty} \frac{1}{(n+1)^2}$$

converges since

$$\sum_{n=1}^{\infty} \frac{1}{n(n+1)}$$

converges. Note that since $\dfrac{1}{(n+1)^2} \geqslant 0$ for all n, convergence is the same as absolute convergence. ■

■ **Example 6.8** Consider now the infinite series $\Sigma_{n=1}^{\infty} \dfrac{(-1)^n}{\sqrt{n}}$. For all $n \geqslant 1$,

$$\left|\frac{(-1)^n}{\sqrt{n}}\right| = \frac{1}{\sqrt{n}} \geq \frac{1}{n}$$

and $\Sigma_{n=1}^{\infty} \frac{1}{n}$ diverges. Hence,

$$\sum_{n=1}^{\infty} \frac{(-1)^n}{\sqrt{n}}$$

does not converge absolutely; but, as we shall see later, it does converge conditionally. ■

The next few results are intended to populate, as painlessly as possible, our stock of examples of convergent infinite series.

6.6 THEOREM Suppose $\{a_n\}_{n=1}^{\infty}$ is a sequence of nonnegative terms such that $a_n \geq a_{n+1}$ for all n. Then $\Sigma_{n=1}^{\infty} a_n$ converges iff $\Sigma_{k=0}^{\infty} 2^k a_{2^k}$ converges.

Proof. Define

$$S_n = a_1 + a_2 + \cdots + a_n$$

and

$$T_k = a_1 + 2a_2 + 4a_4 + \cdots + 2^k a_{2^k}.$$

S_n is the nth partial sum of $\Sigma_{n=1}^{\infty} a_n$, and T_k is the kth partial sum of $\Sigma_{k=0}^{\infty} 2^k a_{2^k}$. Now if $n \leq 2^k$,

$$\begin{aligned} S_n &= a_1 + a_2 + \cdots + a_n \\ &\leq a_1 + (a_2 + a_3) + (a_4 + a_5 + a_6 + a_7) + \cdots + (a_{2^k} + \cdots + a_{2^{k+1}-1}) \\ &\leq a_1 + 2a_2 + 4a_4 + \cdots + 2^k a_{2^k} = T_k \end{aligned}$$

since $a_n \geq a_{n+1}$ for all n.

On the other hand, if $n \geq 2^k$, then

$$\begin{aligned} S_n &= a_1 + a_2 + \cdots + a_n \\ &\geq a_1 + a_2 + (a_3 + a_4) + \cdots + (a_{2^{k-1}+1} + \cdots + a_{2^k}) \\ &\geq \frac{1}{2}a_1 + a_2 + 2a_4 + \cdots + 2^{k-1}a_{2^k} = \frac{1}{2}T_k, \end{aligned}$$

so that $2S_n \geq T_k$.

Now if $\Sigma_{n=1}^{\infty} a_n$ converges, then the sequence $\{S_n\}_{n=1}^{\infty}$ is increasing and it converges to $S = \sup S_n$. Since $2^k \geq 0$ for all k, it suffices to show that $\{T_k\}_{k=1}^{\infty}$ is bounded in order to prove that $\Sigma_{k=0}^{\infty} 2^k a_{2^k}$ converges. Choose k. Then there is n such that $n \geq 2^k$; hence $T_k \leq 2S_n \leq 2S$. Thus, $\{T_k\}_{k=1}^{\infty}$ is bounded, and hence $\Sigma_{k=0}^{\infty} 2^k a_{2^k}$ is convergent.

Suppose now that $\Sigma_{k=0}^{\infty} 2^k a_{2^k}$ is convergent. Then $\{T_k\}_{k=1}^{\infty}$ converges to $T = \sup T_k$. As before, it suffices to show that $\{S_n\}_{n=1}^{\infty}$ is bounded in order to prove that $\Sigma_{n=1}^{\infty} a_n$ converges. Choose n. There is k such that $2^k \geqslant n$. Then $S_n \leqslant T_k \leqslant T$. Thus, $\{S_n\}_{n=1}^{\infty}$ is bounded and hence $\Sigma_{n=1}^{\infty} a_n$ converges.

The infinite series of the form $\Sigma_{n=1}^{\infty} \dfrac{1}{n^p}$, p a real number, are called *p-series*. Theorem 6.6 allows us to classify these quite easily.

6.7 THEOREM $\Sigma_{n=1}^{\infty} \dfrac{1}{n^p}$ converges iff $p > 1$.

Proof. If $p \leqslant 0$, then $\left\{\dfrac{1}{n^p}\right\}_{n=1}^{\infty}$ does not converge to zero, so $\Sigma_{n=1}^{\infty} \dfrac{1}{n^p}$ diverges. If $p > 0$, then $\dfrac{1}{n^p} - \dfrac{1}{(n+1)^p} > 0$ for all n, and Theorem 6.6 may be utilized. Thus, $\Sigma_{n=1}^{\infty} \dfrac{1}{n^p}$ converges iff

$$\sum_{k=0}^{\infty} \frac{2^k}{(2^k)^p}$$

converges. But

$$\frac{2^k}{(2^k)^p} = (2^{1-p})^k,$$

and the series $\Sigma_{k=0}^{\infty} (2^{1-p})^k$ is a geometric series that converges iff $2^{1-p} < 1$, or, in other words, iff $p > 1$.

■ **Example 6.9** Consider the series $\Sigma_{n=1}^{\infty} \dfrac{1}{(2n-1)(2n+1)}$. The nth term of this series may be rewritten as $\dfrac{1}{4n^2 - 1}$ and we want to compare this to a series that is familiar to us. This term almost looks like $\dfrac{1}{4n^2}$ but not quite. We suspect that the series will converge because of Theorem 6.7 with $p = 2$. Follow along:

$$\frac{1}{4n^2 - 1} \leqslant \frac{1}{4n^2 - n^2} \leqslant \frac{1}{3n^2} \quad \text{for all } n \geqslant 1.$$

The series $\Sigma_{n=1}^{\infty} \dfrac{1}{3n^2}$ converges by Theorems 6.7 and 6.3, hence so does $\Sigma_{n=1}^{\infty} \dfrac{1}{4n^2 - 1}$ by Theorem 6.5. Since all terms are positive, the convergence is absolute. ■

6.3 RATIO AND ROOT TESTS

6.8 THEOREM (Ratio Test) Let $\Sigma_{n=1}^{\infty} a_n$ be an infinite series of nonzero terms. Then

1. If there is a real number q such that $0 < q < 1$ and a positive integer N such that $n \geqslant N$ implies
$$\left|\frac{a_{n+1}}{a_n}\right| \leqslant q,$$
then $\Sigma_{n=1}^{\infty} a_n$ converges absolutely.
2. If there is N_0 such that $n \geqslant N_0$ implies
$$\left|\frac{a_{n+1}}{a_n}\right| \geqslant 1,$$
then $\Sigma_{n=1}^{\infty} a_n$ diverges.

Proof. Suppose the hypothesis of (1) holds. Then

$$|a_{N+1}| \leqslant q\ |a_N|,$$
$$|a_{N+2}| \leqslant q\ |a_{N+1}| \leqslant q^2\ |a_N|\ , \ldots, |a_{N+r}| \leqslant q^r\ |a_N|.$$

Thus, for $n \geqslant N$, $|a_n| \leqslant q^n\ |a_N|\ q^{-N}$. Since $0 < q < 1$, $\Sigma_{n=0}^{\infty} q^n$ converges, and

$$\sum_{n=1}^{\infty} |a_N|\ q^{-N} q^n$$

converges. Thus, by the comparison test, $\Sigma_{n=1}^{\infty} a_n$ converges absolutely.

Suppose the hypothesis of (2) holds. Since $|a_n| > 0$ for all n and $\left|\frac{a_{n+1}}{a_n}\right| \geqslant 1$ for $n \geqslant N_0$, we have $|a_n| \leqslant |a_{n+1}|$ for $n \geqslant N_0$; hence, $\{a_n\}_{n=1}^{\infty}$ does not converge to zero. Thus, $\Sigma_{n=1}^{\infty} a_n$ does not converge.

If $\Sigma_{n=1}^{\infty} a_n$ is an infinite series of nonzero terms such that the sequence $\left\{\left|\frac{a_{n+1}}{a_n}\right|\right\}_{n=1}^{\infty}$ is convergent, the next theorem follows from Theorem 6.8. This is the form in which the "ratio test" usually occurs in calculus books.

6.9 THEOREM Suppose $\Sigma_{n=1}^{\infty} a_n$ is an infinite series of nonzero terms such that the sequence $\left\{\left|\frac{a_{n+1}}{a_n}\right|\right\}_{n=1}^{\infty}$ converges to L. Then,

1. If $L < 1$, the series converges absolutely.
2. If $L > 1$, the series diverges.
3. If $L = 1$, no conclusion concerning convergence can be made.

Proof. Suppose $L < 1$. Then $L < \dfrac{L+1}{2} < 1$; and since

$$\left\{\left|\frac{a_{n+1}}{a_n}\right|\right\}_{n=1}^{\infty}$$

converges to L, there is N such that $n \geqslant N$ implies that

$$\left|\frac{a_{n+1}}{a_n}\right| < \frac{L+1}{2}.$$

Hence, by Theorem 6.8, $\Sigma_{n=1}^{\infty} a_n$ converges absolutely.

If $L > 1$, then there is N such that $n \geqslant N$ implies

$$\left|\frac{a_{n+1}}{a_n}\right| \geqslant 1;$$

hence, by Theorem 6.8, $\Sigma_{n=1}^{\infty} a_n$ diverges.

■ **Example 6.10** In order to verify that we gain no information when $L = 1$, we shall give an example of a convergent series for which $L = 1$ and a divergent series for which $L = 1$. Consider first the infinite series $\Sigma_{n=1}^{\infty} \dfrac{1}{n^2}$. By Theorem 6.7 this series converges and

$$\left\{\frac{\dfrac{1}{(n+1)^2}}{\dfrac{1}{n^2}}\right\}_{n=1}^{\infty} = \left\{\frac{n^2}{(n+1)^2}\right\}_{n=1}^{\infty}$$

converges to 1. The series $\Sigma_{n=1}^{\infty} n$ obviously diverges, but again, $\left\{\dfrac{n+1}{n}\right\}_{n=1}^{\infty}$ converges to 1. ■

Theorem 6.9 fails to give any information if

$$\left\{\left|\frac{a_{n+1}}{a_n}\right|\right\}_{n=1}^{\infty}$$

fails to converge. In this case, one may have recourse to Theorem 6.8, but that may also fail to shed any light on the situation. To illustrate this point, consider the infinite series $\Sigma_{n=1}^{\infty} a_n$ where $a_n = \dfrac{1}{3^n}$ for n odd and $a_n = \dfrac{1}{2^n}$ for n even. For n odd,

$$\frac{a_{n+1}}{a_n} = \frac{1}{2}\left(\frac{3}{2}\right)^n;$$

and for n even,

$$\frac{a_{n+1}}{a_n} = \frac{1}{3}\left(\frac{2}{3}\right)^n.$$

Here the sequence $\left\{\frac{a_{n+1}}{a_n}\right\}_{n=1}^{\infty}$ has a subsequence that converges to zero and a subsequence that is unbounded, so neither Theorem 6.8 nor 6.9 applies. The comparison test may be used to verify that this series converges. The "root test," which comes next, is more conclusive than the ratio test in that, if the root test gives no information concerning the convergence of a series, then neither does the ratio test. However, as we shall see by some examples, the root test may be harder to use in some cases. If this were not true, there would be no reason to present the ratio test.

6.10 THEOREM (Root Test) Suppose $\Sigma_{n=1}^{\infty} a_n$ is an infinite series. Then

1. If there is a real number q such that $0 \leqslant q < 1$ and a positive integer N such that $\sqrt[n]{|a_n|} \leqslant q$ for $n \geqslant N$, then $\Sigma_{n=1}^{\infty} a_n$ converges absolutely.
2. If, for infinitely many n, $\sqrt[n]{|a_n|} \geqslant 1$, then $\Sigma_{n=1}^{\infty} a_n$ diverges.

Proof. Suppose there is a real number q such that $0 \leqslant q < 1$ and a positive integer N such that for $n \geqslant N$, $\sqrt[n]{|a_n|} \leqslant q$. Then for $n \geqslant N$, $|a_n| \leqslant q^n$. Since $0 < q < 1$, $\Sigma_{n=1}^{\infty} q^n$ converges; hence, by the comparison test, $\Sigma_{n=1}^{\infty} a_n$ converges absolutely.

Suppose now that $\sqrt[n]{|a_n|} \geqslant 1$ for infinitely many n. Then for infinitely many n, $|a_n| \geqslant 1$, so $\{a_n\}_{n=1}^{\infty}$ does not converge to zero, and $\Sigma_{n=1}^{\infty} a_n$ does not converge.

■ **Example 6.11** Let us now attempt to apply the root test to the earlier example—that is $\Sigma_{n=1}^{\infty} a_n$ where $a_n = \frac{1}{2^n}$ for n even and $a_n = \frac{1}{3^n}$ for n odd. Now $\sqrt[n]{a_n} = \frac{1}{2}$ for n even and $\sqrt[n]{a_n} = \frac{1}{3}$ for n odd. Thus, for all n, $\sqrt[n]{a_n} \leqslant \frac{1}{2}$; hence, by Theorem 6.10, the series converges absolutely. ■

Let us pause to consider a few examples to make use of the root and ratio tests.

■ **Example 6.12** Consider the series

$$\sum_{n=1}^{\infty} \frac{2^n n!}{n^n}.$$

If we attempt to apply the root test, we are faced with considering the sequence whose nth term is

$$\sqrt[n]{\frac{2^n n!}{n^n}} = \frac{2}{n}\sqrt[n]{n!}.$$

Rather than pursue this task further, let us try the ratio test. Now,

$$\frac{a_{n+1}}{a_n} = \frac{2^{n+1}(n+1)!}{(n+1)^{n+1}} \frac{n^n}{2^n n!} = \frac{2}{\left(\frac{n+1}{n}\right)^n} = \frac{2}{\left(1+\frac{1}{n}\right)^n}.$$

In Chapter 1, we observed that the sequence

$$\left\{\left(1+\frac{1}{n}\right)^n\right\}_{n=1}^{\infty}$$

converges with limit e, $2 < e < 3$. Thus,

$$\left\{\frac{2}{\left(1+\frac{1}{n}\right)^n}\right\}_{n=1}^{\infty}$$

converges with limit $\frac{2}{e} < 1$. Therefore, by the ratio test, this series converges. ■

Let $\{a_n\}_{n=0}^{\infty}$ be any sequence of real numbers such that $|a_n| \leqslant n$. Consider $-1 < r < 1$ and the infinite series $\Sigma_{n=0}^{\infty} a_n r^n$. Since little is assumed about the sequence $\{a_n\}_{n=0}^{\infty}$, it seems useless to attempt the ratio test. Let us, rather, attempt to use the root test. Now

$$\sqrt[n]{|a_n|r^n} = \sqrt[n]{|a_n|}\,|r|.$$

We appeal to Theorem 6.10. For each n, $|a_n| \leqslant n$; hence

$$\sqrt[n]{|a_n|} \leqslant \sqrt[n]{n}.$$

The sequence $\{\sqrt[n]{n}\}_{n=1}^{\infty}$ converges to 1; hence, since $-1 < r < 1$, $\{|r|\sqrt[n]{n}\}_{n=1}^{\infty}$ converges to $|r| < 1$. Let

$$q = \frac{|r|+1}{2}.$$

Then $|r| < q < 1$, so there is a positive integer N such that $n \geqslant N$ implies that

$$\sqrt[n]{|a_n|}\,|r| \leqslant \sqrt[n]{n}\,|r| \leqslant q.$$

Thus, by Theorem 6.10, $\Sigma_{n=0}^{\infty} a_n r^n$ converges absolutely. If the readers will review the manner of attack, they should observe that we have proved that $\Sigma_{n=0}^{\infty} nr^n$ converges absolutely for $-1 < r < 1$ and then used the comparison test to show that $\Sigma_{n=0}^{\infty} a_n r^n$ converges absolutely for $-1 < r < 1$ if $|a_n| \leqslant n$.

In summary, the root test has wider scope than the ratio test, although in practice it may be more difficult to use. Neither is subtle with regard to divergence, since both obtain divergence from the fact that $\{a_n\}_{n=1}^{\infty}$ fails to converge to zero. Neither test will serve to identify series that converge conditionally. The next few theorems are of some use in studying such series.

6.4 CONDITIONAL CONVERGENCE

The reader's experience with conditional convergence of series probably relies on the alternating series test. The next two theorems set the stage for that test as well as more general techniques.

6.11 THEOREM Suppose $\{a_n\}_{n=0}^{\infty}$ and $\{b_n\}_{n=0}^{\infty}$ are two sequences of real numbers. Define $A_n = \Sigma_{k=0}^{n} a_k$ for $n \geqslant 0$ and $A_{-1} = 0$. Then if $0 \leqslant p \leqslant q$,

$$\sum_{n=p}^{q} a_n b_n = \sum_{n=p}^{q-1} A_n(b_n - b_{n+1}) + A_q b_q - A_{p-1} b_p .$$

Proof. For $n = 1, 2, \ldots$, we have $a_n = A_n - A_{n-1}$, hence

$$\begin{aligned} \sum_{n=p}^{q} a_n b_n &= \sum_{n=p}^{q} (A_n - A_{n-1}) b_n \\ &= \sum_{n=p}^{q} A_n b_n - \sum_{n=p-1}^{q-1} A_n b_{n+1} \\ &= \sum_{n=p}^{q-1} A_n(b_n - b_{n+1}) + A_q b_q - A_{p-1} b_p . \end{aligned}$$

6.12 THEOREM Assume $\{a_n\}_{n=0}^{\infty}$ and $\{b_n\}_{n=0}^{\infty}$ are sequences of real numbers such that

1. the partial sums of $\Sigma_{n=0}^{\infty} a_n$ are bounded;
2. $b_0 \geqslant b_1 \geqslant b_2 \geqslant \cdots$; and
3. $\{b_n\}_{n=0}^{\infty}$ converges to zero.

Then $\Sigma_{n=1}^{\infty} a_n b_n$ converges.

Proof. As in Theorem 6.11, define $A_n = \Sigma_{k=0}^{n} a_k$ for $n \geqslant 0$ and $A_{-1} = 0$. By (1), $\{A_n\}_{n=1}^{\infty}$ is a bounded sequence; hence there is $M > 0$ such that $|A_n| \leqslant M$ for all n. Choose $\epsilon > 0$. Since $\{b_n\}_{n=1}^{\infty}$ converges to zero, there is N such that $n \geqslant N$ implies $|b_n| < \dfrac{\epsilon}{2M}$. Recall now that $b_n \geqslant 0$ for all n and $b_n - b_{n+1} \geqslant 0$ for all n. Hence, if $N \leqslant p \leqslant q$,

$$\begin{aligned} \left| \sum_{n=p}^{q} a_n b_n \right| &= \left| \sum_{n=p}^{q-1} A_n(b_n - b_{n+1}) + A_q b_q - A_{p-1} b_p \right| \\ &\leqslant \sum_{n=p}^{q-1} |A_n|(b_n - b_{n+1}) + |A_q| b_q + |A_{p-1}| b_p \\ &\leqslant M\left[\sum_{n=p}^{q-1} (b_n - b_{n+1}) + b_q + b_p \right] \\ &= M2b_p < M\frac{2\epsilon}{2M} = \epsilon . \end{aligned}$$

Thus, by Theorem 6.1, $\Sigma_{n=0}^{\infty} a_n b_n$ converges.

As an immediate consequence of Theorem 6.12, we have the "alternating series test."

6.13 THEOREM (Alternating Series Test) Suppose $\{b_n\}_{n=0}^{\infty}$ is a sequence of real numbers such that

1. $\{b_n\}_{n=0}^{\infty}$ converges to zero; and
2. $b_0 \geqslant b_1 \geqslant b_2 \geqslant \cdots$.

Then $\Sigma_{n=0}^{\infty} (-1)^n b_n$ converges.

Proof. Set $a_n = (-1)^n$ for all n. Then

$$A_n = \sum_{k=0}^{n} a_k = 1$$

if n is even and $A_n = 0$ if n is odd. Thus, $\{A_n\}_{n=0}^{\infty}$ is bounded, and we may invoke Theorem 6.12 to conclude that

$$\sum_{n=0}^{\infty} a_n b_n = \sum_{n=0}^{\infty} (-1)^n b_n$$

converges.

In light of Theorem 6.3, one may be tempted to conclude that convergent infinite series may be handled in much the same manner as finite sums. We have postponed the discussion of this problem until now, because an understanding of the distinction between absolute and conditional convergence helps solve the problems to be considered. We have seen that two convergent series may be added term by term, and the resulting series converges to the sum of the two series. Let us now consider multiplication of two infinite series involving what is called the *Cauchy product,* which is motivated by multiplication of finite sums.

DEFINITION Let $\Sigma_{n=0}^{\infty} a_n$ and $\Sigma_{n=0}^{\infty} b_n$ be two infinite series. For each n, define

$$c_n = \sum_{k=0}^{n} a_k b_{n-k}.$$

The infinite series $\Sigma_{n=0}^{\infty} c_n$ is called the *Cauchy product* of the two series $\Sigma_{n=0}^{\infty} a_n$ and $\Sigma_{n=0}^{\infty} b_n$.

This definition may perhaps be best motivated as follows. Suppose that

$$p(x) = \sum_{m=0}^{s} a_m x^m \quad \text{and} \quad q(x) = \sum_{m=0}^{r} b_m x^m$$

are polynomials. Define $a_m = 0$ for $m > s$ and $b_m = 0$ for $m > r$. Then the pointwise product pq is a polynomial and

$$\begin{aligned} p(x)q(x) &= (a_0 + a_1x + \cdots a_sx^s)(b_0 + b_1x + \cdots b_rx^r) \\ &= a_0b_0 + (a_0b_1 + a_1b_0)x + \cdots + (a_0b_n + a_1b_{n-1} + \cdots + a_nb_0)x^n \\ &\quad + \cdots + a_sb_rx^{s+r} \\ &= c_0 + c_1x + \cdots + c_nx^n + \cdots + c_{s+r}x^{s+r} \end{aligned}$$

where $c_n = \sum_{k=0}^{n} a_kb_{n-k}$. In particular, for $x = 1$,

$$(a_0 + a_1 + \cdots + a_s)(b_0 + b_1 + \cdots + b_r) = c_0 + c_1 + \cdots + c_{s+r}.$$

The result that one might hope for now is that the Cauchy product of two convergent series is convergent and converges to the product of the sums of the two series. More precisely, if $\sum_{n=1}^{\infty} a_n$ and $\sum_{n=1}^{\infty} b_n$ converge to A and B, respectively, then $\sum_{n=1}^{\infty} c_n$ converges to AB. It turns out that this is false. In the next few paragraphs, we shall explore these questions and prove the main result.

■ **Example 6.13** Consider the series

$$\sum_{n=0}^{\infty} \frac{(-1)^n}{\sqrt{n+1}}.$$

By the alternating series test, this series converges. We shall now consider the Cauchy product $\sum_{n=0}^{\infty} c_n$ of this series with itself. Now

$$c_n = (-1)^n \sum_{k=0}^{n} \frac{1}{\sqrt{(n-k+1)(k+1)}}.$$

For $n \geqslant k$,

$$(n - k + 1)(k + 1) = \left(\frac{n}{2} + 1\right)^2 - \left(\frac{n}{2} - k\right)^2 \leqslant \left(\frac{n}{2} + 1\right)^2;$$

hence,

$$\frac{1}{\sqrt{(n-k+1)(k+1)}} \geqslant \frac{1}{\frac{n}{2} + 1} = \frac{2}{n+2}.$$

Thus,

$$|c_n| \geqslant \sum_{k=0}^{n} \frac{2}{n+2} = \frac{2(n+1)}{n+2}$$

and $\{c_n\}_{n=0}^{\infty}$ does not converge to zero, and $\sum_{n=0}^{\infty} c_n$ fails to converge. As we shall see by the next theorem, $\sum_{n=0}^{\infty} c_n$ fails to converge because

$$\sum_{n=0}^{\infty} \frac{(-1)^n}{\sqrt{n+1}}$$

converges conditionally. ■

6.14 THEOREM Suppose $\Sigma_{n=0}^{\infty} a_n$ converges absolutely and $\Sigma_{n=0}^{\infty} b_n$ converges with

$$\sum_{n=0}^{\infty} a_n = A \quad \text{and} \quad \sum_{n=0}^{\infty} b_n = B.$$

Define

$$c_n = \sum_{k=0}^{n} a_k b_{n-k}.$$

Then $\Sigma_{n=0}^{\infty} c_n$ converges to AB.

Proof. Define

$$A_n = \sum_{k=0}^{n} a_k, \quad B_n = \sum_{k=0}^{n} b_k,$$
$$C_n = \sum_{k=0}^{n} c_k, \quad \text{and} \quad \beta_n = B_n - B.$$

We wish to prove that $\{C_n\}_{n=1}^{\infty}$ converges to AB. Now for each n,

$$\begin{aligned} C_n &= c_0 + c_1 + c_2 + \cdots + c_n \\ &= a_0b_0 + (a_0b_1 + a_1b_0) + \cdots + (a_0b_n + a_1b_{n-1} + \cdots + a_nb_0) \\ &= a_0B_n + a_1B_{n-1} + \cdots + a_nB_0 \\ &= a_0(B + \beta_n) + a_1(B + \beta_{n-1}) + \cdots + a_n(B + \beta_0). \end{aligned}$$

Define $\gamma_n = a_0\beta_n + a_1\beta_{n-1} + \cdots + a_n\beta_0$. Since $\{A_n\}_{n=0}^{\infty}$ converges to A, the theorem will be proved if we can show that $\{\gamma_n\}_{n=0}^{\infty}$ converges to zero.

Choose $\epsilon > 0$. Let $K = \Sigma_{n=0}^{\infty} |a_n|$ and $M = \sup_{n \geq 0} |\beta_n|$. There is N such that $n \geq N$ implies that

$$|\beta_n| = |B_n - B| < \frac{\epsilon}{2K} \qquad \text{(We may assume } K \neq 0.\text{)}$$

and that for $q \geq N$,

$$\sum_{p=N+1}^{q} |a_p| < \frac{\epsilon}{2M}. \qquad \text{(We may assume } M \neq 0.\text{)}$$

Thus, for $n \geq 2N$,

$$\begin{aligned} |\gamma_n| &= |a_0\beta_n + a_1\beta_{n-1} + \cdots + a_n\beta_0| \leq \sum_{k=0}^{n} |a_k\beta_{n-k}| \\ &= \sum_{k=0}^{N} |a_k\beta_{n-k}| + \sum_{k=N+1}^{n} |a_k\beta_{n-k}| \\ &\leq \frac{\epsilon}{2K}\sum_{k=0}^{N} |a_k| + M\sum_{k=N+1}^{n} |a_k| < \frac{\epsilon}{2K}K + M\frac{\epsilon}{2M} = \epsilon. \end{aligned}$$

A related theorem in which no absolute convergence is assumed is needed to round out the picture. We shall state it without proof. See the project at the end of Chapter 7 for a proof.

6.15 THEOREM Suppose $\Sigma_{n=0}^{\infty} a_n$ and $\Sigma_{n=0}^{\infty} b_n$ converge to A and B, respectively. Define

$$c_n = \sum_{k=0}^{n} a_k b_{n-k}.$$

If $\Sigma_{n=0}^{\infty} c_n$ converges to C, then $C = AB$.

For finite sums, the order in which the terms are arranged has no effect upon the sum. We shall now concern ourselves with the similar situation for infinite series.

DEFINITION Let $\Sigma_{n=0}^{\infty} a_n$ be an infinite series. If T is any 1–1 function from $\{0, 1, 2, \ldots\}$ onto $\{0, 1, 2, \ldots\}$, then the infinite series $\Sigma_{n=0}^{\infty} a_{T(n)}$ is called a *rearrangement* of $\Sigma_{n=0}^{\infty} a_n$.

Of course, we are now concerned with questions about the convergence of the rearrangements of a convergent infinite series and what the sums of such rearrangements might be. Our experience up to now leads us to suspect that all is well with absolutely convergent series, so let us look at a conditionally convergent series. Consider the conditionally convergent series

$$\sum_{n=1}^{\infty} (-1)^{n+1} \frac{1}{n}$$

and denote its sum by S. (S happens to be approximately 0.693; actually, it is log 2, but that will have to wait until Chapter 7.) Then

$$\sum_{n=1}^{\infty} (-1)^{n+1} \frac{1}{2n}$$

converges to $\frac{S}{2}$. We now consider a new series $\Sigma_{n=1}^{\infty} a_n$ where $a_n = 0$ if n is odd and

$$a_n = \frac{(-1)^{(n+2)/2}}{n}$$

if n is even. $\Sigma_{n=1}^{\infty} a_n$ converges to $\frac{S}{2}$, since all we have done is insert a zero into the series $\Sigma_{n=1}^{\infty} (-1)^{n+1} \frac{1}{2n}$ between each pair of terms. To see that this is the case, let S_n be the nth partial sum of $\Sigma_{n=1}^{\infty} (-1)^{n+1} \frac{1}{2n}$, and let A_n be the nth partial sum of $\Sigma_{n=1}^{\infty} a_n$. Then $A_1 = 0$, $A_{2n} = S_n = A_{2n+1}$ for all $n \geqslant 1$, hence $\{A_n\}_{n=1}^{\infty}$ converges to the same limit as $\{S_n\}_{n=1}^{\infty}$, which is $\dfrac{S}{2}$. Now adding

$$\sum_{n=1}^{\infty} (-1)^{n+1} \frac{1}{n} \quad \text{and} \quad \sum_{n=1}^{\infty} a_n,$$

we obtain a series that converges to $\dfrac{3S}{2}$. Deleting the zeros from this resulting series, it is easy to see that it is a rearrangement of $\Sigma_{n=1}^{\infty} (-1)^{n+1} \dfrac{1}{n}$. To clarify this situation, let us write out a few terms of each series in question.

$$\sum_{n=1}^{\infty} (-1)^{n+1} \frac{1}{n} = 1 - \frac{1}{2} + \frac{1}{3} - \frac{1}{4} + \frac{1}{5} - \frac{1}{6} + \frac{1}{7} - \frac{1}{8} + \cdots$$

$$\sum_{n=1}^{\infty} a_n = 0 + \frac{1}{2} + 0 - \frac{1}{4} + 0 + \frac{1}{6} + 0 - \frac{1}{8} + 0 + \cdots$$

$$\sum_{n=1}^{\infty} \left[(-1)^{n+1} \frac{1}{n} + a_n\right] = 1 + 0 + \frac{1}{3} - \frac{1}{2} + \frac{1}{5} + 0 + \frac{1}{7} - \frac{1}{4} + \cdots$$

$$= 1 + \frac{1}{3} - \frac{1}{2} + \frac{1}{5} + \frac{1}{7} - \frac{1}{4} + \cdots$$

Thus, we have found a rearrangement of $\Sigma_{n=1}^{\infty} \dfrac{(-1)^{n+1}}{n}$ that converges to a different limit. Of course,

$$\sum_{n=1}^{\infty} \frac{(-1)^{n+1}}{n}$$

is only conditionally convergent. Let us now attack the problem for absolutely convergent series. If a series converges absolutely, it is easy to see that any rearrangement will converge absolutely, since all one needs to do to verify convergence for series with nonnegative terms is to show that the partial sums are bounded. We seek to solve the more difficult problem of showing that all rearrangements converge to the same limit.

6.16 THEOREM Let $\Sigma_{n=1}^{\infty} a_n$ be an absolutely convergent series converging to A and $\Sigma_{n=1}^{\infty} a_{T(n)}$ be any rearrangement of $\Sigma_{n=1}^{\infty} a_n$. Then

$$\sum_{n=1}^{\infty} a_{T(n)}$$

converges to A.

Proof. For each n, let

$$S_n = \sum_{k=1}^{n} a_k \quad \text{and} \quad S_n^1 = \sum_{k=1}^{n} a_{T(k)}.$$

Now $\{S_n\}_{n=1}^\infty$ converges to A. We seek to prove that $\{S_k^1\}_{k=1}^\infty$ also converges to A. Choose $\epsilon > 0$. There is N such that for $n \geqslant N$,

$$|S_n - A| = \left| \sum_{k=n+1}^{\infty} a_k \right| \leqslant \sum_{k=n+1}^{\infty} |a_k| < \frac{\epsilon}{2}$$

(since $\Sigma_{n=1}^\infty a_n$ converges absolutely). Now T is a 1–1 function mapping $\{1, 2, \ldots\}$ onto $\{1, 2, \ldots\}$; hence, there is $K \geqslant N$ such that $\{1, 2, 3, \ldots, N\} \subset \{T(1), T(2), \ldots, T(K)\}$. Assume $n \geqslant K$. Then

$$|S_n^1 - A| \leqslant |S_n^1 - S_N| + |S_N - A| < \left| \sum_{k=1}^{n} a_{T(k)} - \sum_{k=1}^{N} a_k \right| + \frac{\epsilon}{2}$$

$$\leqslant \sum_{k=N+1}^{\infty} |a_k| + \frac{\epsilon}{2} < \frac{\epsilon}{2} + \frac{\epsilon}{2}.$$

Thus, $\{S_n^1\}_{n=1}^\infty$ converges to A, and the theorem is proved.

Theorem 6.16 has a converse in a sense. If $\Sigma_{n=1}^\infty a_n$ is an infinite series with the property that every rearrangement converges to the same sum, the series is absolutely convergent. The next few pages will be devoted to this result. But a few preliminary remarks are needed.

DEFINITION Let $\{a_n\}_{n=1}^\infty$ be any sequence of real numbers. For each n, define $a_n^+ = a_n$ if $a_n \geqslant 0$ and $a_n^+ = 0$ if $a_n < 0$. Define $a_n^- = a_n$ if $a_n < 0$ and $a_n^- = 0$ if $a_n \geqslant 0$. Thus, if

$$\{a_n\}_{n=1}^\infty = \left\{1, -\frac{1}{2}, \frac{1}{3}, -\frac{1}{4}, \frac{1}{5}, -\frac{1}{6}, \ldots\right\},$$

then

$$\{a_n^+\}_{n=1}^\infty = \left\{1, 0, \frac{1}{3}, 0, \frac{1}{5}, \ldots\right\}$$

and

$$\{a_n^-\}_{n=1}^\infty = \left\{0, -\frac{1}{2}, 0, -\frac{1}{4}, 0, -\frac{1}{6}, \ldots\right\}.$$

In any case, $a_n = a_n^+ + a_n^-$ for all n.

Suppose now that $\Sigma_{n=1}^\infty a_n$ converges absolutely. Then for each n,

$$|a_n^+| \leqslant |a_n| \quad \text{and} \quad |a_n^-| \leqslant |a_n|;$$

hence, $\Sigma_{n=1}^\infty a_n^+$ and $\Sigma_{n=1}^\infty a_n^-$ converge absolutely. Conversely, if $\Sigma_{n=1}^\infty a_n^+$ and $\Sigma_{n=1}^\infty (-a_n^-)$ converge (actually, this is absolute convergence since both are series of nonnegative terms), then, since $|a_n| = a_n^+ - a_n^-$,

$$\sum_{n=1}^{\infty} |a_n|$$

converges by Theorem 6.3. This result proves the following theorem.

6.17 THEOREM The infinite series $\Sigma_{n=1}^{\infty} a_n$ converges absolutely iff the series $\Sigma_{n=1}^{\infty} a_n^+$ and $\Sigma_{n=1}^{\infty} a_n^-$ converge. If $\Sigma_{n=1}^{\infty} a_n^+$ converges with sum A^+ and $\Sigma_{n=1}^{\infty} a_n^-$ converges with sum A^-, then the sum of $\Sigma_{n=1}^{\infty} a_n$ is $A^+ + A^-$, and the sum of $\Sigma_{n=1}^{\infty} |a_n|$ is $A^+ - A^-$.

To recap our results, if a series $\Sigma_{n=1}^{\infty} a_n$ converges conditionally, then at least one of the two series, $\Sigma_{n=1}^{\infty} a_n^+$ or $\Sigma_{n=1}^{\infty} a_n^-$, diverges. Since we may write

$$a_n^+ = a_n - a_n^- \quad \text{and} \quad a_n^- = a_n - a_n^+,$$

we see that, if $\Sigma_{n=1}^{\infty} a_n$ converges conditionally, then both $\Sigma_{n=1}^{\infty} a_n^+$ and $\Sigma_{n=1}^{\infty} a_n^-$ must diverge. Since $a_n^+ \geq 0$, the partial sums of $\Sigma_{n=1}^{\infty} a_n^+$ form an increasing sequence unbounded from above; and, since $a_n^- \leq 0$, the partial sums of $\Sigma_{n=1}^{\infty} a_n^-$ must form a decreasing sequence unbounded from below. When we consider rearrangements of series, we shall have recourse to this result. If we denote the nth partial sum of

$$\sum_{n=1}^{\infty} a_n, \quad \sum_{n=1}^{\infty} a_n^+, \quad \text{and} \quad \sum_{n=1}^{\infty} a_n^-$$

by S_n, S_n^+, and S_n^-, respectively, then

$$\{S_n\}_{n=1}^{\infty} = \{S_n^+ + S_n^-\}_{n=1}^{\infty}$$

converges, whereas $\{S_n^+\}_{n=1}^{\infty}$ and $\{S_n^-\}_{n=1}^{\infty}$ do not. The manner in which the cancellation takes place in the series $\Sigma_{n=1}^{\infty} a_n$ may be very complicated, and our techniques for recognizing conditionally convergent series will unfortunately be applicable only to certain special types.

We are now equipped to state and prove a partial converse of Theorem 6.16.

6.18 THEOREM If $\Sigma_{n=1}^{\infty} a_n$ converges conditionally, then, given any real number r, there is a rearrangement of $\Sigma_{n=1}^{\infty} a_n$ that converges to r.

Proof. First of all, notice that $\{S_n^+\}_{n=1}^{\infty}$ and $\{S_n^-\}_{n=1}^{\infty}$ are unbounded sequences and $\{a_n\}_{n=1}^{\infty}$ converges to zero. Let n_1 be the first positive integer such that

$$a_1^+ + a_2^+ + \cdots + a_{n_1}^+ > r.$$

There is such n_1 since $\{S_n^+\}_{n=1}^{\infty}$ is an increasing sequence not bounded from above. Next, let m_1 be the least positive integer such that

$$(a_1^+ + a_2^+ + \cdots + a_{n_1}^+) + (a_1^- + a_2^- + \cdots + a_{m_1}^-) < r.$$

As before, such an m_1 exists since $\{S_n^-\}_{n=1}^{\infty}$ is a decreasing sequence bounded from below. Now let n_2 be the least positive integer such that

$$(a_1^+ + \cdots + a_{n_1}^+) + (a_1^- + \cdots + a_{m_1}^-) + (a_{n_1+1}^+ + \cdots + a_{n_2}^+) > r.$$

As argued before, such an n_2 exists. Now choose m_2 to be the least positive integer such that

$$(a_1^+ + \cdots + a_{n_1}^+) + (a_1^- + \cdots + a_{m_1}^-) \\ + (a_{n_1+1}^+ + \cdots + a_{n_2}^+) + (a_{m_1+1}^- + \cdots + a_{m_2}^-) < r.$$

Again, such an m_2 exists. Continuing in this way, deleting zeros when they occur, we obtain a rearrangement of the given series. Moreover, the rearrangement converges to r since $\{a_n\}_{n=1}^{\infty}$ converges to zero.

The proof of Theorem 6.18 leaves quite a few details to the reader to verify. However, by this time we assume the reader is sophisticated enough to fill in those details. The intent here is to provide the basic idea that makes the proof work.

6.5 POWER SERIES

We are now going to consider a special type of infinite series called *power series.* Since the reader has encountered this topic in the study of calculus, many of our results will not be new. However, armed with a firm grasp of the basic facts about infinite series, we should be prepared to gain a much fuller understanding of the inner workings of power series.

DEFINITION Let $\{a_n\}_{n=1}^{\infty}$ be a sequence of real numbers. For each real number x, we may consider the infinite series

$$\sum_{n=0}^{\infty} a_n x^n.$$

We shall refer to this collection of infinite series as the *power series* generated by $\{a_n\}_{n=0}^{\infty}$ or, more briefly, as the power series $\Sigma_{n=0}^{\infty} a_n x^n$.

Given a power series $\Sigma_{n=0}^{\infty} a_n x^n$, we find that for some values of x, it converges (for $x = 0$, for example), whereas for some other values of x, it may diverge. Our first considerations will be to find the set of points where a given power series converges and determine the general nature of that set.

■ **Example 6.14** Let us consider several familiar examples. We have already seen that $\Sigma_{n=0}^{\infty} x^n$ converges for $-1 < x < 1$ and diverges elsewhere. On the set $(-1, 1)$,

$$\sum_{n=0}^{\infty} x^n = \frac{1}{1 - x}.$$

This is, of course, the power series generated by the sequence $\{a_n\}_{n=0}^{\infty}$, with $a_n = 1$ for all n. The power series $\Sigma_{n=0}^{\infty} \dfrac{x^n}{n!}$ is also familiar; it converges to e^x for all real x. (We shall prove this later.) ■

Intuitively, an infinite series will converge if the nth term tends to zero "fast enough." Thus, one is led to suspect that if $\Sigma_{n=0}^{\infty} a_n x^n$ converges and $|y| < x$, then maybe $\Sigma_{n=0}^{\infty} a_n y^n$ converges. However, our past experiences with conditionally convergent series should cause us to proceed with caution.

6.19 THEOREM Let $\Sigma_{n=0}^{\infty} a_n x^n$ be a power series that converges for $x = x_0$ and diverges for $x = x_1$. Then

1. $\Sigma_{n=0}^{\infty} a_n x^n$ converges absolutely for $|x| < |x_0|$; and
2. $\Sigma_{n=0}^{\infty} a_n x^n$ diverges for $|x| > |x_1|$.

Proof. Assume $\Sigma_{n=0}^{\infty} a_n x_0^n$ converges and $|x| < |x_0|$ with $x_0 \neq 0$. Since $\Sigma_{n=0}^{\infty} a_n x_0^n$ converges, the sequence $\{a_n x_0^n\}_{n=0}^{\infty}$ converges to zero, hence is bounded; that is, there is $M \in R$ such that $|a_n x^n| \leqslant M$ for all n. For all n,

$$|a_n x^n| = |a_n x_0^n| \cdot \left|\frac{x}{x_0}\right|^n \leqslant M \left|\frac{x}{x_0}\right|^n.$$

Since

$$\left|\frac{x}{x_0}\right| < 1,$$

the series

$$\sum_{n=0}^{\infty} M \left|\frac{x}{x_0}\right|^n$$

converges, and hence, by the comparison test, $\Sigma_{n=0}^{\infty} a_n x^n$ converges absolutely.

If $|x| > |x_1|$ and $\Sigma_{n=0}^{\infty} a_n x^n$ converges, then by (1), $\Sigma_{n=0}^{\infty} a_n x_1^n$ converges absolutely, contrary to assumption. Hence, $\Sigma_{n=0}^{\infty} a_n x^n$ diverges.

Let us summarize these results. Suppose $\Sigma_{n=0}^{\infty} a_n x^n$ is a power series and let $C = \{x : \Sigma_{n=0}^{\infty} a_n x^n \text{ converges}\}$. The possibilities for C are listed below:

i. $C = R$, in which case $\Sigma_{n=0}^{\infty} a_n x^n$ converges for all $x \in R$.
ii. $C = \{0\}$, in which case $\Sigma_{n=0}^{\infty} a_n x^n$ diverges for all $x \neq 0$.
iii. There is $r > 0$ such that $(-r, r) \subset C \subset [-r, r]$, in which case $\Sigma_{n=0}^{\infty} a_n x^n$ converges absolutely for $|x| < r$ and diverges for $|x| > r$.

Cases (i) and (ii) are self-explanatory. We should justify (iii). Suppose $C \neq \{0\}$ and $C \neq R$. Since $C \neq R$, there is $x_1 \in R$ such that $\Sigma_{n=0}^{\infty} a_n x_1^n$ diverges. Hence, by Theorem 6.19, for each $x \in C$, $|x| \leqslant |x_1|$. Therefore, C is bounded; let $r = \sup C$. If $|x| > r$, then $x \not\in C$, so $\Sigma_{n=0}^{\infty} a_n x^n$ diverges. If $|x| < r$, then there is a member p of C such that $|x| < p \leqslant r$ since $r = \sup C$. Thus, by Theorem 6.19, $\Sigma_{n=0}^{\infty} a_n x^n$ converges absolutely, since $\Sigma_{n=0}^{\infty} a_n p^n$ converges. Since $C \neq \{0\}$, there is $x \in C$ such that $x \neq 0$; hence, $r \geqslant |x| > 0$. We cannot assert what happens at r or at $-r$.

■ **Example 6.15** Consider again the power series $\Sigma_{n=0}^{\infty} \dfrac{x^n}{n!}$. For each n,

$$\frac{x^{n+1}}{(n+1)!} \cdot \frac{n!}{x^n} = \frac{x}{n+1};$$

hence, by the ratio test, this series converges for all x, as stated earlier. Similarly, the ratio test may be used to show that $\Sigma_{n=0}^{\infty} n!x^n$ converges only for $x = 0$. The geometric series $\Sigma_{n=0}^{\infty} x^n$ converges for $-1 < x < 1$ and diverges for $|x| \geqslant 1$. The power series $\Sigma_{n=0}^{\infty} \dfrac{x^n}{n}$ converges for $-1 \leqslant x < 1$ and diverges for both $|x| > 1$ and $x = 1$. ■

If $\Sigma_{n=0}^{\infty} a_n x^n$ is a power series, then the set of points at which the series converges is either the set of all real numbers, $\{0\}$, or an interval of positive finite length centered at zero that may contain all, none, or one of its endpoints. We choose to consider $\{0\}$ as an interval of zero radius and R as an interval of infinite radius. With this convention, we feel free to speak of the *interval of convergence* of a power series. In particular, $\Sigma_{n=0}^{\infty} a_n x^n$ converges absolutely at any point of its interval of convergence that is not an endpoint of that interval. If the interval of convergence is R, the series converges absolutely at each point; and if $(-r, r) \subset C \subset [-r, r]$, where C is the interval of convergence, then the series converges absolutely at x for $x \in (-r, r)$. As shown by our examples, the behavior at the endpoints is unpredictable.

If $\Sigma_{n=0}^{\infty} a_n x^n$ has an interval of convergence C that is different from R and $\{0\}$, then there is a unique real number r such that

$$(-r, r) \subset C \subset [-r, r].$$

This number r is called the *radius of convergence* of the power series. We now seek a way to determine r. Since we have already used the ratio test in one case, let's see what we can deduce from this test.

Suppose $\Sigma_{n=0}^{\infty} a_n x^n$ is a power series with $a_n \neq 0$ for all n and $x \neq 0$. Consider the ratio

$$\left|\frac{a_{n+1}x^{n+1}}{a_n x^n}\right| = \left|\frac{a_{n+1}}{a_n}\right| \cdot |x|.$$

For the ratio test, we may refer to either Theorem 6.8 or 6.9. For the simplest case, let us assume that the sequence

$$\left\{\left|\frac{a_{n+1}}{a_n}\right|\right\}_{n=1}^{\infty}$$

converges to $L \neq 0$. Then the sequence

$$\left\{\left|\frac{a_{n+1}x^{n+1}}{a_n x^n}\right|\right\}_{n=1}^{\infty}$$

converges to $x \cdot L$. Hence, $\Sigma_{n=0}^{\infty} a_n x^n$ converges for $|x| \cdot L < 1$, or for $|x| < \frac{1}{L}$, and diverges for $|x| \cdot L > 1$, or for $|x| > \frac{1}{L}$. Thus, the radius of convergence is

$\frac{1}{L}$ if $L \neq 0$. If $L = 0$, then $|x| \cdot L = 0 < 1$ for all x, so the series converges for all x. We summarize these results as follows:

6.20 THEOREM Suppose $\Sigma_{n=0}^{\infty} a_n x^n$ is a power series with $a_n \neq 0$ for all n such that

$$\left\{\left|\frac{a_{n+1}}{a_n}\right|\right\}_{n=0}^{\infty}$$

converges to L. Then

1. if $L = 0$, the series converges for all x; and
2. if $L \neq 0$, $\frac{1}{L}$ is the radius of convergence.

Theorem 6.20 has the obvious advantage of being easy to apply when it is applicable. However, when

$$\left\{\left|\frac{a_{n+1}}{a_n}\right|\right\}_{n=0}^{\infty}$$

fails to converge, Theorem 6.20 yields no information. At this point, we still have recourse to Theorem 6.8.

6.21 THEOREM Let $\Sigma_{n=0}^{\infty} a_n x^n$ be a power series with $a_n \neq 0$ for all n. Then

1. If there are real numbers $q > 0$ and N such that $n \geqslant N$ implies that
$$\left|\frac{a_{n+1}}{a_n}\right| \leqslant q,$$
then $\Sigma_{n=0}^{\infty} a_n x^n$ converges absolutely for $|x| < \dfrac{1}{q}$;
2. If there are real numbers $p > 0$ and N such that for all $n \geqslant N$,
$$\left|\frac{a_{n+1}}{a_n}\right| \geqslant p,$$
then $\Sigma_{n=0}^{\infty} a_n x^n$ converges absolutely for $|x| > \dfrac{1}{p}$.

Proof. Suppose the condition of (1) holds and $|x| < \dfrac{1}{q}$. Then by (1) of Theorem 6.8, $\Sigma_{n=0}^{\infty} a_n x^n$ converges absolutely.

Suppose now the condition of (2) holds and $|x| > \dfrac{1}{p}$. Then by (2) of Theorem 6.8, $\Sigma_{n=0}^{\infty} a_n x^n$ diverges.

Theorem 6.21 tells us that the radius of convergence r satisfies $\dfrac{1}{q} \leqslant r \leqslant \dfrac{1}{p}$ for all p and q that satisfy the hypotheses of this theorem. If

$$\left\{\left|\frac{a_{n+1}}{a_n}\right|\right\}_{n=1}^{\infty}$$

converges to $L \neq 0$, then p may be chosen to be any number less than L and q any number greater than L, hence this yields again the fact that $\frac{1}{L} \leqslant r \leqslant \frac{1}{L}$, which means $r = \frac{1}{L}$.

■ **Example 6.16** Theorem 6.21 is an inefficient device for finding the radius of convergence of a power series since it is conclusive only if

$$\left\{\left|\frac{a_{n+1}}{a_n}\right|\right\}_{n=0}^{\infty}$$

converges, in which case one should apply Theorem 6.20. Let us consider the power series $\Sigma_{n=0}^{\infty} a_n x^n$ where $a_n = 2$ for n odd and $a_n = 3$ for n even. In this case,

$$\frac{a_{n+1}}{a_n} = \frac{2}{3}$$

if n is even, and

$$\frac{a_{n+1}}{a_n} = \frac{3}{2}$$

if n is odd. Thus, Theorem 6.21 tells us only that the radius of convergence is less than or equal to $\frac{3}{2}$ and greater than or equal to $\frac{2}{3}$. In this case, it is easy to see that the radius of convergence is equal to 1. If we let $a_n = 2^{-n}$ for n even and $a_n = 3^{-n}$ for n odd, then

$$\frac{a_{n+1}}{a_n} = \left(\frac{2}{3}\right)^n \left(\frac{1}{3}\right)$$

for n even and

$$\frac{a_{n+1}}{a_n} = \left(\frac{3}{2}\right)^n \left(\frac{1}{2}\right)$$

for n odd. For this power series, there are no real numbers q that satisfy (1) of Theorem 6.21 and no real numbers p that satisfy (2) of Theorem 6.21, so Theorem 6.21 is useless. The reader is invited to show that the radius of convergence of this power series is 2. ■

Perhaps the root test may yield better results, although we suspect it may be more difficult to apply. Let us experiment briefly before we formulate a theorem. If there is a real number q and a positive integer N such that $n \geqslant N$ implies that $\sqrt[n]{|a_n|} \leqslant q$, then

$$\sqrt[n]{|a_n x^n|} = \sqrt[n]{|a_n|} \cdot |x| \leqslant q \cdot |x|;$$

hence, by Theorem 6.10, the series $\Sigma_{n=0}^{\infty} a_n x^n$ converges absolutely for $|x| < \frac{1}{q}$ if $q \neq 0$ and for all x if $q = 0$. Thus, the radius of convergence r satisfies $r \geqslant \frac{1}{q}$ if $q \neq 0$. Since there may be many possible choices for q, we would like to prove a more conclusive theorem. Our concern centers around the possible candidates for q.

Let $\Sigma_{n=0}^{\infty} a_n x^n$ be a power series, and define $A = \{q :$ there is N such that $n \geqslant N$ implies that $\sqrt[n]{|a_n|} \leqslant q\}$. Clearly, zero is a lower bound for A, and, if the sequence $\{\sqrt[n]{|a_n|}\}_{n=1}^{\infty}$ is bounded, then A is not empty. Let us suppose A is nonempty, and define $b = \inf A$. Since the radius of convergence r of the power series $\Sigma_{n=0}^{\infty} a_n x^n$ satisfies $r \geqslant \frac{1}{q}$ for all $q \in A$, then, if $b \neq 0$, we might suspect that $r \geqslant \frac{1}{b}$; in fact, we hope $r = \frac{1}{b}$. To prove this, we need to know some of the properties of b; this notion might be worth considering in a more general setting.

Suppose $\{b_n\}_{n=1}^{\infty}$ is any bounded sequence of real numbers, and define $A = \{q :$ for some N, $n \geqslant N$ implies that $b_n \leqslant q\}$. Since $\{b_n\}_{n=1}^{\infty}$ is bounded, A is nonvoid and also bounded from below. Let $b = \inf A$. Consider any real numbers x and y such that $x < b < y$. There is $q \in A$ such that $b \leqslant q < y$ since $b = \inf A$, so there is N such that $n \geqslant N$ implies $b_n \leqslant q < y$. Since $x < b$, $x \notin A$, then for infinitely many n, $b_n > x$. Thus, we have infinitely many members of the sequence greater than x, whereas only finitely many are larger than y. Since x and y are any real numbers satisfying $x < b < y$, we are led to suspect that there is a subsequence of $\{b_n\}_{n=1}^{\infty}$ that converges to b. Moreover, no number larger than b has this property. We are now ready for a theorem.

6.22 THEOREM Let $\{b_n\}_{n=1}^{\infty}$ be a bounded sequence of real numbers, and define

$$A = \{p : \text{there is a subsequence of } \{b_n\}_{n=1}^{\infty} \text{ converging to } p\}$$

and

$$B = \{q : \text{there is } N \text{ such that for } n \geqslant N,\ b_n \leqslant q\}.$$

Then the following are equivalent:

1. $b = \inf B$.
2. $b = \sup A$ and $b \in A$.
3. For each $\epsilon > 0$, there is N such that $n \geqslant N$ implies that $b_n < b + \epsilon$, and for each M, there is $m \geqslant M$ such that $b_m > b - \epsilon$.

Proof. Assume (1) holds. Choose $\epsilon > 0$. Then $b + \epsilon$ is not a lower bound for B and $b - \epsilon \notin B$ since $b - \epsilon < b$. Thus, there is $q \in B$ such that $b \leqslant q < b + \epsilon$; hence, there is N such that $n \geqslant N$ implies that $b_n \leqslant q < b + \epsilon$. Since $b - \epsilon$ is not in B, there must be infinitely many terms of the sequence that satisfy $b_n > b - \epsilon$. Consequently, (1) implies (3).

Assume (3) holds. Then for $\epsilon_k = \frac{1}{k}$, with k any positive integer, there is N_k such that $n \geqslant N_k$ implies

$$b_n < b + \epsilon_k,$$

and there is $n_k \geqslant N_k$ such that $b - \epsilon_k < b_{n_k}$. Thus,

$$|b_{n_k} - b| < \frac{1}{k}$$

for each k, so $\{b_{n_k}\}_{k=1}^{\infty}$ converges to b and $b \in A$. Consider any $x > b$. There is N such that $n \geqslant N$ implies that

$$b_n < \frac{x + b}{2} < x,$$

so there is a neighborhood of x that contains only finitely many terms of the sequence $\{b_n\}_{n=1}^{\infty}$. This means that no subsequence of $\{b_n\}_{n=1}^{\infty}$ can converge to x, or $x \notin A$. Thus, we have $b \in A$ and for all $x \in A$, $x \leqslant b$. We have shown that (3) implies (2).

Assume (2) holds. We have proved that (1) implies (3) and (3) implies (2); to finish the proof, we need only prove that (2) implies (1). Suppose $q \in B$. Then there is N such that $n \geqslant N$ implies $b_n \leqslant q$. Since there are but finitely many terms of the sequence $\{b_n\}_{n=1}^{\infty}$ that are larger than q, there can be no subsequence of $\{b_n\}_{n=1}^{\infty}$ converging to any x larger than q. This means that $b \leqslant q$ since $b \in A$. Therefore, b is a lower bound for B; it remains to be shown that b is the greatest lower bound of B. Suppose $b < a$ and a is a lower bound for B. Then $\frac{a + b}{2}$ does not belong to B, and so we may choose a sequence of positive integers $\{n_k\}_{k=1}^{\infty}$ such that $n_1 < n_2 < \cdots$ and such that $b_{n_k} > \frac{a + b}{2}$. Since the sequence $\{b_{n_k}\}_{k=1}^{\infty}$ is bounded, it has a convergent subsequence, and the limit of that subsequence, which we will call L, must be greater than or equal to $\frac{a + b}{2}$. This convergent subsequence is also a subsequence of the sequence $\{b_n\}_{n=1}^{\infty}$; hence, $L \in A$. But we had assumed $b = \sup A$ and $b < \frac{a + b}{2} \leqslant L$. Thus, any real number $a > b$ cannot be a lower bound for B. We conclude that $b = \inf B$. Therefore, (2) implies (1).

Let us make a few observations about Theorem 6.22. First of all, the boundedness of $\{b_n\}_{n=1}^{\infty}$ implies that B is nonvoid and bounded, so there is a unique real number b satisfying (1) and hence (2) and (3). Condition (1) is important here in that this description of b was the motivating factor in the investigation of the properties of b. Condition (2) describes b as the largest subsequential limit of the sequence, and the theorem guarantees that there is a largest subsequential limit if the sequence is bounded. Moreover, if $\{b_n\}_{n=1}^{\infty}$ converges, then it converges to b. It seems natural that one might be able to prove a similar theorem concerning $\inf\{p :$ some subsequence of $\{b_n\}_{n=1}^{\infty}$ converges to $p\}$. See the exercises for some interesting challenges along these lines.

DEFINITION Let $\{b_n\}_{n=1}^{\infty}$ be a sequence of real numbers, and define $B = \{p :$ there is a subsequence of $\{b_n\}_{n=1}^{\infty}$ converging to $p\}$. If B is not void and if $\{b_n\}_{n=1}^{\infty}$ is bounded from above, define

$$\limsup_{n\to\infty} b_n = \sup B.$$

If B is not void and if $\{b_n\}_{n=1}^{\infty}$ is bounded from below, define

$$\liminf_{n\to\infty} b_n = \inf B.$$

■ **Example 6.17** Consider the sequences $\{b_n\}_{n=1}^{\infty}$, where $b_n = (-1)^n$. Then

$$\limsup_{n\to\infty} b_n = 1 \quad \text{and} \quad \liminf_{n\to\infty} b_n = -1.$$

If $b_n = n$ for each n, B is void; hence, we define neither $\limsup_{n\to\infty} b_n$ nor $\liminf_{n\to\infty} b_n$. If $b_n = n$ for n even and $b_n = 1 - \dfrac{1}{n}$ for n odd, then $\liminf_{n\to\infty} b_n = 1$, whereas $\limsup_{n\to\infty} b_n$ is not defined. We have proved that the set of all rational numbers is countable, so we may arrange the rational numbers in $(0, 1)$ in a sequence $\{r_n\}_{n=1}^{\infty}$. The reader is urged to prove that in this case

$$B = [0, 1], \quad \liminf_{n\to\infty} r_n = 0, \quad \text{and} \quad \limsup_{n\to\infty} r_n = 1. \qquad ■$$

We now return to the problem that led us to this latest diversion.

6.23 THEOREM Let $\Sigma_{n=0}^{\infty} a_n x^n$ be a power series.

1. If the sequence $\{\sqrt[n]{|a_n|}\}_{n=0}^{\infty}$ is unbounded, $\Sigma_{n=0}^{\infty} a_n x^n$ converges only for $x = 0$.
2. If the sequence $\{\sqrt[n]{|a_n|}\}_{n=0}^{\infty}$ converges to zero, then $\Sigma_{n=0}^{\infty} a_n x^n$ converges for all x.
3. If the sequence $\{\sqrt[n]{|a_n|}\}_{n=0}^{\infty}$ is bounded and

$$a = \limsup_{n\to\infty} \sqrt[n]{|a_n|} \neq 0,$$

then $\frac{1}{a}$ is the radius of convergence.

Proof. 1. Choose $x \neq 0$. For each N, there is $n \geq N$ such that

$$\sqrt[n]{|a_n|} > \frac{1}{|x|}.$$

Hence, $|a_n x^n| > 1$ for some $n \geq N$. In particular, $\{a_n x^n\}_{n=0}^{\infty}$ does not converge to zero for $x \neq 0$, hence $\Sigma_{n=0}^{\infty} a_n x^n$ diverges for $x \neq 0$.

2. Suppose $\{\sqrt[n]{|a_n|}\}_{n=0}^{\infty}$ converges to zero and $x \neq 0$. Define $\epsilon = \frac{1}{|2x|}$. There is N such that $n \geqslant N$ implies that

$$\sqrt[n]{|a_n|} < \epsilon = \frac{1}{|2x|}.$$

Thus, for $n \geqslant N$,

$$|a_n x^n| < \epsilon^n |x|^n = \left(\frac{1}{2}\right)^n.$$

Therefore, by the comparison test, $\Sigma_{n=0}^{\infty} a_n x^n$ converges for all x.

3. Suppose that

$$a = \limsup_{n \to \infty} \sqrt[n]{|a_n|} \neq 0.$$

Consider any $x \neq 0$ such that $|x| < \frac{1}{a}$. Then $a < \frac{1}{|x|}$ and, by (1) of Theorem 6.22, there are real numbers q and N such that $a \leqslant q < \frac{1}{x}$ and for $n \geqslant N$, $\sqrt[n]{|a_n|} \leqslant q < \frac{1}{x}$. Hence, for $n \geqslant N$,

$$\sqrt[n]{|a_n x^n|} \leqslant q|x| < 1.$$

Therefore, by Theorem 6.10, $\Sigma_{n=0}^{\infty} a_n x^n$ converges absolutely. To show that $\frac{1}{a}$ is actually the radius of convergence, we must show that the power series $\Sigma_{n=0}^{\infty} a_n x^n$ diverges for $|x| > \frac{1}{a}$. Suppose $|x| > \frac{1}{a}$. Then $a > \frac{1}{|x|}$, and hence, by (3) of Theorem 6.22, there are infinitely many n such that

$$\sqrt[n]{|a_n|} > \frac{1}{|x|}.$$

So for infinitely many n, $|a_n x^n| > 1$; hence $\{a_n x^n\}_{n=1}^{\infty}$ does not converge to zero. Thus, for $|x| > \frac{1}{a}$, the series $\Sigma_{n=0}^{\infty} a_n x^n$ diverges.

We have as an immediate corollary to Theorem 6.23 a slightly different statement of Theorem 6.10.

6.24 THEOREM Let $\Sigma_{n=0}^{\infty} a_n$ be an infinite series. Then

1. If $\{\sqrt[n]{|a_n|}\}_{n=0}^{\infty}$ is unbounded, the series diverges.
2. If $\{\sqrt[n]{|a_n|}\}_{n=0}^{\infty}$ is bounded, then $\Sigma_{n=0}^{\infty} a_n$ converges absolutely if

$$\limsup_{n \to \infty} \sqrt[n]{|a_n|} < 1$$

and diverges if

$$\limsup_{n\to\infty} \sqrt[n]{|a_n|} > 1.$$

Proof. The series $\Sigma_{n=0}^{\infty} a_n$ converges iff the power series $\Sigma_{n=0}^{\infty} a_n x^n$ converges at $x = 1$. The theorem then follows directly from Theorem 6.23.

The reader will find it instructive to restate Theorems 6.21 and 6.8 in terms of the new concepts just introduced.

■ **Example 6.18** Let us reconsider an example mentioned earlier. Define $a_n = 2^{-n}$ for n odd and $a_n = 3^{-n}$ for n even. Then

$$\sqrt[n]{|a_n|} = \frac{1}{2}$$

for n odd and

$$\sqrt[n]{|a_n|} = \frac{1}{3}$$

for n even. Hence, $\{\sqrt[n]{|a_n|}\}_{n=1}^{\infty}$ has precisely two subsequential limits, $\frac{1}{2}$ and $\frac{1}{3}$. Since the larger of the two is $\limsup_{n\to\infty} \sqrt[n]{|a_n|}$, the radius of convergence is 2, as claimed earlier. ■

6.6 TAYLOR SERIES

We now take up a slightly different aspect of power series. Recall that for all n and $-1 < x < 1$,

$$1 + x + x^2 + \cdots + x^n = \frac{1 - x^{n+1}}{1 - x}.$$

Since $\{x^{n+1}\}_{n=1}^{\infty}$ converges to zero for $-1 < x < 1$, the power series $\Sigma_{n=0}^{\infty} x^n$ converges to $\dfrac{1}{1 - x}$ for $-1 < x < 1$. We shall now inquire what conditions on a function f will guarantee that there is a power series $\Sigma_{n=0}^{\infty} a_n x^n$ such that for all x in some interval, $\Sigma_{n=0}^{\infty} a_n x^n$ converges to $f(x)$. Since the point zero is not magic in this discussion, we shall now extend our previous results by a few simple remarks. The function $g(x) = \dfrac{1}{x}$ is not well behaved at zero, but we may still write

$$g(x) = \frac{1}{x} = \frac{-1}{1 - (x + 1)} = -\sum_{n=0}^{\infty} (x + 1)^n$$

for $|x + 1| < 1$ or $-2 < x < 0$. By the same sort of device, we may also write

$$g(x) = \frac{\frac{1}{5}}{1 - \left[-\frac{(x-5)}{5}\right]} = \sum_{n=0}^{\infty} \frac{(-1)^n}{5^{n+1}}(x-5)^n$$

for $\left|\frac{x-5}{5}\right| < 1$, or for $0 < x < 10$. If a power series $\Sigma_{n=0}^{\infty} a_n y^n$ converges for $-r < y < r$, then the power series $\Sigma_{n=0}^{\infty} a_n(x-a)^n$ converges for $-r < x - a < r$ and, hence, for $a - r < x < a + r$. Thus, the set of points where $\Sigma_{n=0}^{\infty} a_n(x-a)^n$ converges is either $\{a\}$, the set of all real numbers, or an interval of positive length centered at a. (The same remarks as before hold concerning the endpoints.)

Suppose $f: [a, b] \to R$ is differentiable on (a, b) and continuous on $[a, b]$. Then by the Mean-Value Theorem,

$$f(b) = f(a) + f'(c)(b - a)$$

for some $c \in (a, b)$. The next result, Taylor's Theorem, is an extension of the Mean-Value Theorem. You may recall some versions of this theorem in Chapter 5.

6.25 TAYLOR'S THEOREM Suppose that $f: [a, b] \to R$ is n-times differentiable on $[a, b]$ and that $f^{(n)}$ is continuous on $[a, b]$ and differentiable on (a, b). Assume $x_0 \in [a, b]$. Then for each $x \in [a, b]$ with $x \neq x_0$, there is c between x and x_0 such that

$$f(x) = f(x_0) + \sum_{k=1}^{n} \frac{f^{(k)}(x_0)}{k!}(x - x_0)^k + \frac{f^{(n+1)}(c)}{(n+1)!}(x - x_0)^{n+1}.$$

Proof. Define

$$F(t) = f(t) + \sum_{k=1}^{n} \frac{f^{(k)}(t)}{k!}(x - t)^k + M(x - t)^{n+1},$$

where M is chosen so that $F(x_0) = f(x)$. This is possible since $x - x_0 \neq 0$. Now F is continuous on $[a, b]$ and differentiable on (a, b), and

$$F(x) = f(x) = F(x_0);$$

hence, by Rolle's Theorem, there is c between x and x_0 such that

$$0 = F'(c) = \frac{f^{(n+1)}(c)}{n!}(x - c)^n - (n + 1)M(x - c)^n.$$

Thus,

$$M = \frac{f^{(n+1)}(c)}{(n+1)!}.$$

Hence,

$$\begin{aligned} f(x) &= F(x_0) \\ &= f(x_0) + \sum_{k=1}^{n} \frac{f^{(k)}(x_0)}{k!}(x - x_0)^k + \frac{f^{(n+1)}(c)}{(n + 1)!}(x - x_0)^{n+1}, \end{aligned}$$

as the theorem states.

Assume the hypotheses of Theorem 6.25 and fix $x_0 \in [a, b]$. Define

$$p(x) = f(x_0) + \sum_{k=1}^{n} \frac{f^{(k)}(x_0)}{k!}(x - x_0)^k.$$

Now p is a polynomial of degree less than or equal to n, and Taylor's formula gives us a means of approximating the error when we approximate f by the polynomial p.

■ **Example 6.19** Let us illustrate Taylor's Theorem with a rather easy example. We again assume knowledge of the function $f(x) = e^x$. Now, f is differentiable everywhere and $f'(x) = e^x = f(x)$. Hence, f has continuous derivatives of all orders; namely $f^{(n)} = f$ for all n. Then for each x and each n, there is c_n between x and 0 such that

$$\begin{aligned} e^x = f(x) &= f(0) + \sum_{k=1}^{n} \frac{f^{(k)}(0)}{k!} x^k + \frac{f^{(n+1)}(c_n)}{(n + 1)} x^{n+1} \\ &= 1 + \sum_{k=1}^{n} \frac{x^k}{k!} + e^{c_n} \frac{x^{n+1}}{(n + 1)!}. \end{aligned}$$

Since f is increasing, $e^{-|x|} \leqslant e^c \leqslant e^{|x|}$ for all c between x and 0. Thus,

$$\left\{ e^{c_n} \frac{x^{n+1}}{(n + 1)!} \right\}_{n=1}^{\infty}$$

converges to zero, and so

$$\left\{ 1 + \sum_{k=1}^{n} \frac{x^k}{k!} \right\}_{n=1}^{\infty}$$

converges to e^x. As x was arbitrary, we have

$$\sum_{k=0}^{\infty} \frac{x^k}{k!} = e^x$$

for all x. ■

■ **Example 6.20** Assuming the basic facts about the functions $g(x) = \sin x$ and $h(x) = \cos x$, we know that g is differentiable everywhere, $g'(x) = \cos x$, which is also differentiable everywhere, and $g''(x) = -\sin x = -g(x)$. Thus, g has continuous derivatives of all orders, and $|g^{(n)}(x)| \leqslant 1$ for all n and all x. As in the preceding example, this guarantees that

$$\sum_{n=0}^{\infty} \frac{g^{(n)}(0)}{n!} x^n$$

converges to $\sin x$ for all real x. Since $\cos 0 = 1$ and $\sin 0 = 0$, we have

$$\sin x = \sum_{n=0}^{\infty} \frac{(-1)^n}{(2n+1)!} x^{2n+1}$$

for all x. A similar line of reasoning allows us to conclude that

$$\cos x = \sum_{n=0}^{\infty} \frac{(-1)^n}{(2n)!} x^{2n}$$

for all x. ■

The interested reader is referred to Chapter 7 for more on power series and the convergence of sequences of functions.

■ **Example 6.21** Taylor's Theorem is also useful for writing polynomials in powers of $x - a$ where $a \neq 0$. If f is a polynomial, then, for some n, $f^{(n)}(x) = 0$ for all x and the series is a finite sum. To illustrate, we shall write the polynomial $p(x) = 2x^3 - 4x^2 - x + 1$ in powers of $x - 1$. First, we compute the derivatives and evaluate at $x = 1$.

$$\begin{aligned} p(x) &= 2x^3 - 4x^2 - x + 1 & p(1) &= -2 \\ p'(x) &= 6x^2 - 8x - 1 & p'(1) &= -3 \\ p''(x) &= 12x - 8 & p''(1) &= 4 \\ p'''(x) &= 12 & p'''(1) &= 12 \end{aligned}$$

The fourth derivative of p is 0. So for all x,

$$\begin{aligned} p(x) &= p(1) + p'(1)(x-1) + \frac{p''(x)}{2!}(x-1)^2 + \frac{p'''(1)}{3!}(x-1)^3 \\ &= -2 - 3(x-1) + 2(x-1)^2 + 2(x-1)^3. \end{aligned}$$ ■

EXERCISES

6.1 CONVERGENCE OF INFINITE SERIES

1. Let $\{a_n\}_{n=1}^{\infty}$ be a sequence of real numbers. Prove that

$$\sum_{n=1}^{\infty} (a_n - a_{n+1})$$

converges iff $\{a_n\}_{n=1}^{\infty}$ converges. If $\Sigma_{n=1}^{\infty} (a_n - a_{n+1})$ converges, what is the sum?

2. Let $\Sigma_{n=1}^{\infty} a_n$ converge. Let $\{n_k\}_{k=1}^{\infty}$ be a subsequence of the sequence of positive integers. For each k, define

$$b_k = a_{nk-1+1} + \cdots + a_{nk} \qquad \text{where } n_0 = 0.$$

Prove that $\Sigma_{k=1}^{\infty} b_k$ converges and that

$$\sum_{k=1}^{\infty} b_k = \sum_{n=1}^{\infty} a_n.$$

3. Prove that $\Sigma_{n=1}^{\infty} 2^n r^n$ converges if $|r| < \frac{1}{2}$ and find the sum.

4. Prove that the series $\Sigma_{n=0}^{\infty} 3^{-n}$ converges and find the limit.

5. Determine whether the series $\Sigma_{n=1}^{\infty} (\sqrt{n+1} - \sqrt{n})$ converges or diverges. Justify your conclusion.

6. Use induction to show that $1 + \frac{1}{\sqrt{2}} + \cdots + \frac{1}{\sqrt{n}} \geq \sqrt{n}$ for $n \geq 1$. Can you use this fact to determine whether the series $\Sigma_{n=1}^{\infty} \frac{1}{\sqrt{n}}$ converges or diverges?

7. Use induction to show that $2\left(1 + \frac{1}{8} + \frac{1}{27} + \cdots + \frac{1}{n^3}\right) < 3 - \frac{1}{n^2}$ for $n \geq 2$. Does the series $\Sigma_{n=1}^{\infty} \frac{1}{n^3}$ converge? Justify your conclusions.

8. Write an infinite series for the repeating decimal for the rational number $\frac{5}{9}$ and prove that it converges to $\frac{5}{9}$.

9. Find the rational number that is the limit of the repeating decimal $0.\overline{15}$.

10. Prove Theorem 6.3.

11. Prove that the series $\Sigma_{n=2}^{\infty} \left(\frac{5}{2^n} - \frac{3}{5^n}\right)$ converges and find the sum.

12. Show that the series $\Sigma_{n=1}^{\infty} \left(\frac{2}{n} + \frac{1}{2^n}\right)$ diverges. Use Theorem 6.3.

6.2 ABSOLUTE CONVERGENCE AND THE COMPARISON TEST

13. Suppose $\Sigma_{n=1}^{\infty} a_n$ converges absolutely and $\{b_n\}_{n=1}^{\infty}$ is bounded. Prove that $\Sigma_{n=1}^{\infty} a_n b_n$ converges absolutely.

14. If $\Sigma_{n=1}^{\infty} a_n$ converges absolutely, prove that $\Sigma_{n=1}^{\infty} a_n^2$ converges. Is the converse true? Is the statement true if $\Sigma_{n=1}^{\infty} a_n$ converges conditionally?

15. Determine which of the following infinite series converge.

a. $\displaystyle\sum_{n=1}^{\infty} \frac{1}{(n+1)(2n-1)}$ b. $\displaystyle\sum_{n=1}^{\infty} \frac{n}{n+1}$ c. $\displaystyle\sum_{k=1}^{\infty} \frac{2}{3k}$

d. $\displaystyle\sum_{m=1}^{\infty} \frac{\sqrt{m+1} - \sqrt{m}}{m}$ e. $\displaystyle\sum_{m=1}^{\infty} \frac{3^m}{5^{m+1}}$ f. $\displaystyle\sum_{m=1}^{\infty} \frac{1}{m^2}\left(\frac{100m+1}{m}\right)$

16. (Limit comparison test). Prove the following generalization of Theorem 6.5. Suppose $\Sigma_{n=0}^{\infty} a_n$ and $\Sigma_{n=0}^{\infty} b_n$ are series of positive terms such that

$$\left\{\frac{a_n}{b_n}\right\}_{n=1}^{\infty}$$

converges to $L \neq 0$. Then $\Sigma_{n=0}^{\infty} a_n$ and $\Sigma_{n=0}^{\infty} b_n$ either both diverge or both converge. What can be concluded if $L = 0$?

17. Apply the limit comparison test (Exercise 16) to the following series:

a. $\sum_{n=1}^{\infty} \frac{\sqrt{n} + 17}{n^2 - 64n - 112}$ b. $\sum_{n=1}^{\infty} \frac{1}{\sqrt{n^2 + 64n} + \sqrt{n^2 + 3}}$

6.3 RATIO AND ROOT TESTS

18. Give an example of an infinite series for which Theorem 6.8 yields a conclusion, but Theorem 6.9 does not.

19. Use Theorem 6.9 to determine the values of r for which $\Sigma_{n=0}^{\infty} nr^n$ converges.

20. Prove that $\{nx^n\}_{n=1}^{\infty}$ converges to zero if $|x| < 1$.

21. Prove the following version of the root test. If $\Sigma_{n=1}^{\infty} a_n$ is an infinite series such that $\{\sqrt[n]{|a_n|}\}_{n=1}^{\infty}$ converges to L, then, (1) if $L < 1$, the series converges absolutely; and (2) if $L > 1$, the series diverges.

22. If $\{a_n\}_{n=1}^{\infty}$ is a sequence of positive real numbers such that

$$\left\{\frac{a_{n+1}}{a_n}\right\}_{n=1}^{\infty}$$

converges to L, prove that $\{\sqrt[n]{a_n}\}_{n=1}^{\infty}$ converges to L.

23. Prove that

$$\left\{\frac{\sqrt[n]{n!}}{n}\right\}_{n=1}^{\infty}$$

converges, and find the limit. You might want to look at Example 6.12.

24. Test the following series for convergence:

a. $\sum_{n=1}^{\infty} n^p p^n,\ p > 0$ b. $\sum_{n=1}^{\infty} (\sqrt[n]{n} - 1)^n$

c. $\sum_{n=1}^{\infty} n^{-1-(1/n)}$ d. $\sum_{n=2}^{\infty} \frac{1}{p^n - q^n},\ 0 < q < p$

25. Determine those values of p for which $\Sigma_{n=2}^{\infty} \frac{1}{n(\log n)^p}$ converges.

26. For each positive integer n, define

$$\gamma_n = 1 + \frac{1}{2} + \cdots + \frac{1}{n} - \log n.$$

Prove that $\{\gamma_n\}_{n=1}^{\infty}$ converges. (Use the fact that $\log x = \int_1^x \frac{dt}{t}$ for $x > 0$.)

6.4 CONDITIONAL CONVERGENCE

27. If $\Sigma_{n=1}^{\infty} a_n$ converges and $\{b_n\}_{n=1}^{\infty}$ is monotone and bounded, prove that $\Sigma_{n=1}^{\infty} a_n b_n$ converges.

28. Suppose $\{a_n\}_{n=1}^{\infty}$ is a sequence of positive real numbers converging to zero such that $a_n \geqslant a_{n+1}$ for all n. Then by the alternating series test, $\Sigma_{n=1}^{\infty} (-1)^n a_n$ converges; call the sum S. Let S_n be the nth partial sum of $\Sigma_{n=1}^{\infty} (-1)^n a_n$. Prove that $|S_n - S| \leqslant a_{n+1}$.

29. Let $\Sigma_{n=1}^{\infty} a_n$ be an infinite series and $\{n_k\}_{k=1}^{\infty}$ a subsequence of the sequence of positive integers. Prove that if $\Sigma_{n=1}^{\infty} a_n$ converges absolutely, then $\Sigma_{k=1}^{\infty} a_{n_k}$ converges absolutely. What can be concluded if $\Sigma_{n=1}^{\infty} a_n$ converges conditionally?

30. Prove that $\Sigma_{k=1}^{\infty} \sin \dfrac{kx}{k}$ converges. (*Hint:* Use Theorem 6.12.)

31. Test each of the following series for absolute convergence, conditional convergence, or divergence:

a. $\displaystyle\sum_{n=1}^{\infty} \frac{(-1)^n n}{n^2 - 5n + 1}$ b. $\displaystyle\sum_{n=1}^{\infty} \frac{(-1)^n (n-5)}{n^3 - 7n - 9}$

c. $\displaystyle\sum_{n=1}^{\infty} (\sqrt{n+1} - \sqrt{n}) a_n$ where $a_n = (-1)^{f(n)}$ and $f(n) = \left[\dfrac{n}{2}\right]$; that is, $f(n)$ is the integer part of $\dfrac{n}{2}$.

d. $\displaystyle\sum_{n=1}^{\infty} (-1)^n \frac{n}{n+1}$

6.5 POWER SERIES

32. Suppose that $\Sigma_{n=0}^{\infty} a_n$ diverges and that $\{a_n\}_{n=0}^{\infty}$ is bounded. Prove that the radius of convergence of $\Sigma_{n=0}^{\infty} a_n x^n$ is equal to 1.

33. Suppose $\Sigma_{n=1}^{\infty} a_n$ converges conditionally. Prove that the radius of convergence of $\Sigma_{n=1}^{\infty} a_n x^n$ is equal to 1.

34. Show that $\Sigma_{n=1}^{\infty} n! x^n$ converges only for $x = 0$.

35. Show that $\Sigma_{n=1}^{\infty} \dfrac{x^n}{n}$ converges iff $-1 \leq x < 1$.

36. Determine the radius of convergence of the power series $\Sigma_{n=1}^{\infty} \dfrac{2^n n!}{n^n} x^n$.

37. Determine the interval of convergence of the power series $\Sigma_{n=1}^{\infty} \dfrac{1}{n^p} x^n$ for different values of p.

38. Show that the power series $\Sigma_{n=0}^{\infty} a_n x^n$ and $\Sigma_{n=1}^{\infty} n a_n x^{n-1}$ either both converge for all x, both converge only for $x = 0$, or both have the same finite nonzero radius of convergence.

39. Let $\{a_n\}_{n=1}^{\infty}$ be a sequence of real numbers bounded from below and let $A = \{p :$ there is a subsequence of $\{a_n\}_{n=1}^{\infty}$ converging to $p\}$. Suppose A is nonvoid. Define $a = \inf A$. Prove that $a \in A$ and that, for each $\epsilon > 0$ there is N such that for all $n \geq N$, $a - \epsilon < a_n$ and there are infintely many m such that $a_m < a + \epsilon$.

40. State and prove theorems similar to Theorems 6.8 and 6.21 in terms of

$$\liminf_{n \to \infty} \left| \frac{a_{n+1}}{a_n} \right| \quad \text{and} \quad \limsup_{n \to \infty} \left| \frac{a_{n+1}}{a_n} \right|.$$

41. Find the interval of convergence of the power series $\Sigma_{n=1}^{\infty} a_n x^n$ where a_n is as given below. Be sure to check the endpoints.

a. $a_n = (n+1)(n+2)$ b. $a_n = \sin n$

c. $a_n = 3^{-n}\sqrt{n}$

d. $a_n = \dfrac{(n!)^2}{(2n)!}$

e. $a_n = \left(1 + \dfrac{1}{n}\right)^{n2}$

f. $a_n = \dfrac{3^n}{n} + \dfrac{2^n}{n^2}$

6.6 TAYLOR SERIES

42. Write the Taylor series for $\log x$ using powers of $x - 1$. See the project at the end of the last chapter if you have forgotten the derivative of $\log x$. Prove that this series converges to $\log x$ for $1 \leqslant x < 2$.

43. Write the Taylor series for $\sqrt{1 - x}$ in powers of x. Prove that the series converges to $\sqrt{1 - x}$ for $-1 < x \leqslant 0$.

44. If $x > 0$, show that $\left|(1 + x)^{1/3} - \left(1 + \dfrac{x}{3} - \dfrac{x^2}{9}\right)\right| \leqslant \dfrac{5x^3}{81}$.

45. Show that for $|x| \leqslant 1$, $\left|\sin x - \left(x - \dfrac{x^3}{6} + \dfrac{x^5}{120}\right)\right| \leqslant \dfrac{1}{5040}$.

46. Use the Taylor series for e^x to prove that $1 + x \leqslant e^x$ for all $x > 0$.

47. Suppose that $a_1, a_2, \ldots, a_n$ are positive real numbers, and let $A = \dfrac{1}{n}\Sigma_{k=1}^n a_k$. If we let $x = \dfrac{a_k}{A} - 1$, the inequality in Exercise 46 shows that $\dfrac{a_k}{A} \leqslant e^{(a_k/A)-1}$. Multiply these inequalities together to prove that $(a_1 a_2 \cdot \cdots \cdot a_n)^{1/n} \leqslant \dfrac{a_1 + a_2 + \cdots + a_n}{n}$.

48. Define $f(x) = e^{-1/x^2}$ if $x \neq 0$ and $f(0) = 0$. Show that f is infinitely differentiable (that is, has derivatives of all orders), but if $x \neq 0$,

$$f(x) \neq \sum_{n=1}^{\infty} \frac{f^{(n)}(0)}{n!} x^n.$$

PROJECT

The purpose of this project is to develop Wallis' formula.

For $n = 0, 1, 2, \ldots$, define $I_n = \displaystyle\int_0^{\pi/2} \sin^n t\, dt$.

1. Show that $I_{n+1} \leqslant I_n$ for $n > 0$.
2. Show by integration by parts, that $I_n = \dfrac{n-1}{n} I_{n-2}$ for all $n \geqslant 2$.
3. Prove that $\dfrac{I_{2n+1}}{I_{2n-1}} = \dfrac{2n}{2n+1}$.
4. Prove that $\dfrac{2n}{2n+1} = \dfrac{I_{2n+1}}{I_{2n-1}} \leqslant \dfrac{I_{2n+1}}{I_{2n}} \leqslant 1$ for all $n \geqslant 1$.
5. Conclude that $\left\{\dfrac{I_{2n+1}}{I_{2n}}\right\}_{n=1}^{\infty}$ converges to 1.

6. Prove that $\dfrac{I_{2n+1}}{I_{2n}} = \dfrac{2\cdot 2\cdot 4\cdot 4\cdot 6\cdot 6\cdot\cdots(2n)(2n)\cdot 2}{1\cdot 3\cdot 3\cdot 5\cdot 5\cdot 7\cdot 7\cdot\cdots(2n-1)(2n-1)(2n+1)\pi}$.

Parts 5 and 6 above yield Wallis' Formula:

$$\lim_{n\to\infty}\frac{2\cdot 2\cdot 4\cdot 4\cdot 6\cdot 6\cdot\cdots(2n)(2n)}{1\cdot 3\cdot 3\cdot 5\cdot 5\cdot 7\cdot 7\cdot\cdots(2n-1)(2n-1)(2n+1)} = \frac{\pi}{2}$$

Wallis' formula gives $\frac{\pi}{2}$ as an *infinite product,* defined as the limit of partial products, in much the same way that we defined the infinite sum as the limit of partial sums. If you continue your study of analysis—in particular, analysis of complex variables—you will learn more about infinite products.

CHAPTER 7

Sequences and Series of Functions

In Chapter 6, we considered power series but left many rather obvious questions unasked and unanswered. In this chapter we seek out these questions and answer some of them.

Suppose $\sum_{n=0}^{\infty} a_n x^n$ is a power series that converges for $-r < x < r$ where $r > 0$. For each integer $n \geqslant 0$, define S_n: $(-r, r) \to R$ by

$$S_n(x) = \sum_{k=0}^{n} a_k x^k$$

for all $x \in (-r, r)$. Therefore, for each $x \in (-r, r)$ and each integer $n \geqslant 0$, $S_n(x)$ is the nth partial sum of the convergent series $\sum_{n=0}^{\infty} a_n x^n$, so the sequence $\{S_n(x)\}_{n=0}^{\infty}$ converges to $\sum_{n=0}^{\infty} a_n x^n$. It is natural to define a function S: $(-r, r) \to R$ by defining $S(x)$ to be the limit of the convergent sequence $\{S_n(x)\}_{n=0}^{\infty}$ or, equivalently,

$$S(x) = \sum_{n=0}^{\infty} a_n x^n,$$

for all $x \in (-r, r)$. It is customary to denote the limit of a convergent sequence $\{a_n\}_{n=1}^{\infty}$ by $\lim_{n \to 1} a_n$. We have avoided this usage because of possible conflict with the notation in Chapter 2. However, by now we should not be disturbed by this notation. Thus, we write

$$S(x) = \lim_{n \to \infty} S_n(x)$$

for each $x \in (-r, r)$. Each S_n is a polynomial and, hence, is continuous everywhere, differentiable everywhere, and Riemann-integrable on every interval of finite length. Does S inherit any of these nice properties? If S is differentiable at $x \in (-r, r)$, is it true that

$$S'(x) = \lim_{n \to \infty} S_n'(x) = \sum_{n=1}^{\infty} n a_n x^{n-1}?$$

If S is Riemann-integrable on $[a, b] \subset (-r, r)$, is it true that

$$\int_a^b S(t)\,dt = \lim_{n\to\infty} \int_a^b S_n(t)\,dt = \sum_{n=0}^{\infty} a_n \left[\frac{b^{n+1} - a^{n+1}}{n+1}\right]?$$

Perhaps there lurks in the reader's past some experience with Fourier series—that is, series of the type

$$\sum_{n=0}^{\infty} (a_n \cos nx + b_n \sin nx).$$

Here again, define $T_n\colon R \to R$ by

$$T_n(x) = \sum_{k=0}^{n} (a_k \cos kx + b_k \sin kx)$$

for each integer $n \geqslant 0$. If this series converges for all $x \in E$, one may define a function $T\colon E \to R$ by

$$T(x) = \lim_{n\to\infty} T_n(x) = \sum_{n=0}^{\infty} (a_n \cos nx + b_n \sin nx)$$

for all $x \in E$. Again, the same questions may be asked concerning the nature of T. These last remarks lead us to suspect that it may be more profitable to consider the situation in general rather than concentrate our attention on power series alone.

7.1 POINTWISE AND UNIFORM CONVERGENCE

DEFINITION Suppose $\{f_n\}_{n=0}^{\infty}$ is a sequence of functions and E is a subset of R such that $E \subset \operatorname{dom} f_n$ for each integer $n \geqslant 0$. Then $\{f_n\}_{n=0}^{\infty}$ *converges pointwise* on E if for each $x \in E$ the sequence

$$\{f_n(x)\}_{n=0}^{\infty}$$

converges. If $\{f_n\}_{n=0}^{\infty}$ converges pointwise on E, define $f\colon E \to R$ by

$$f(x) = \lim_{n\to\infty} f_n(x)$$

for each $x \in E$.

We say that $\{f_n\}_{n=0}^{\infty}$ converges pointwise to f on E. It is common practice to call f the *limit function* of the sequence $\{f_n\}_{n=0}^{\infty}$ and write $f = \lim_{n\to\infty} f_n$.

■ **Example 7.1** As a simple and familiar example, for each integer $n \geqslant 0$, define $f_n\colon (-1, 1) \to R$ by

$$f_n(x) = \sum_{k=0}^{n} x^k$$

for each $x \in (-1, 1)$. Then $\{f_n\}_{n=1}^{\infty}$ converges pointwise on $(-1, 1)$ to the function $g(x) = \dfrac{1}{1-x}$. ■

We depart from the realm of power series to seek other examples.

■ **Example 7.2** Define for each integer $n > 0$, f_n: $[0, 1] \to R$ by

$$f_n(x) = x^n$$

for each $x \in [0, 1]$. For $x \in [0, 1)$,

$$\{f_n(x)\}_{n=1}^{\infty}$$

converges to zero and for $x = 1$,

$$\{f_n(x)\}_{n=1}^{\infty}$$

converges to 1. Thus, the limit function is not continuous although each f_n is continuous. ■

■ **Example 7.3** For each positive integer n, define g_n: $[0, 1] \to R$ by

$$g_n(0) = 0$$

$$g_n(x) = n \quad \text{for } 0 < x \leq \frac{1}{n} \qquad \text{and}$$

$$g_n(x) = 0 \quad \text{for } \frac{1}{n} < x \leq 1.$$

Then $\{g_n\}_{n=1}^{\infty}$ converges pointwise on $[0, 1]$ to the function g, which is zero for all $x \in [0, 1]$. Each g_n is Riemann-integrable on $[0, 1]$, as is g, but

$$\left\{ \int_0^1 g_n(t)\, dt \right\}_{n=1}^{\infty}$$

converges to 1, whereas $\int_0^1 g(t)\, dt = 0$. ■

■ **Example 7.4** Consider the sequence $\{h_n\}_{n=1}^{\infty}$ defined by

$$h_n(x) = \frac{1}{x} \quad \text{for } \frac{1}{n} \leq x < 1 \qquad \text{and}$$

$$h_n(x) = n^2 x \quad \text{for } 0 \leq x < \frac{1}{n}.$$

The sequence $\{h_n(x)\}_{n=1}^{\infty}$ converges to $\frac{1}{x}$ for $0 < x < 1$ and converges to zero at $x = 0$. Here each h_n is continuous, but the limit function is unbounded and, hence, not Riemann-integrable. Note further that the sequence

$$\left\{ \int_0^1 h_n(t)\, dt \right\}_{n=1}^{\infty}$$

is unbounded. ■

■ **Example 7.5** The sequence $\{f_n\}_{n=1}^{\infty}$ given above in Example 7.2 yields an example of a sequence of differentiable functions that converges to a function that fails to be differentiable. We now present a sequence of differentiable functions $\{k_n\}_{n=1}^{\infty}$ such that the limit function k is differentiable, but for some x,

$$k'(x) \neq \lim_{n\to\infty} k_n'(x).$$

To this end, define, for each positive integer n, k_n: $[-1, 1] \to R$ by

$$k_n(x) = \frac{x}{1 + nx^2}$$

for all $x \in [-1, 1]$. For each positive integer n, k_n is differentiable on $[-1, 1]$ and

$$k_n'(x) = \frac{1 - nx^2}{(1 + nx^2)^2}.$$

From this, it is easy to see that the minimum and maximum values of k_n occur at $\dfrac{-1}{\sqrt{n}}$ and $\dfrac{1}{\sqrt{n}}$, respectively. Hence,

$$|k_n(x)| \leqslant \frac{1}{2\sqrt{n}}$$

for all $x \in [-1, 1]$ and all positive integers n. It follows quite easily that $\{k_n(x)\}_{n=1}^{\infty}$ converges to zero for all $x \in [-1, 1]$. The limit function is thus constant on $[-1, 1]$ and hence differentiable, with the derivative equal to zero everywhere on $[-1, 1]$. However $k_n'(0) = 1$ for all n. ■

Rather than be discouraged by this list of examples, we shall examine these examples more carefully with the hope of finding a substitute for pointwise convergence that yields affirmative answers to some of the questions posed earlier.

Suppose $\{f_n\}_{n=1}^{\infty}$ converges pointwise to a function f on a set E. This means that for each $\epsilon > 0$ and each $x \in E$, there is N (probably depending on both x and ϵ), such that for each positive integer n, $n \geqslant N$ implies that

$$|f_n(x) - f(x)| < \epsilon.$$

Again, for each positive integer n, define $f_n(x) = x^n$ for each $x \in [0, 1]$. Given $0 < x < 1$ and $\epsilon > 0$, there is a lower bound for the possible values of N; in fact, N must be larger than $\left|\dfrac{\log \epsilon}{\log x}\right|$. For x close to 1 but different from 1, N must be quite large, for, given any $0 < \epsilon < 1$ and any positive integer m, there is $0 < x < 1$ such that $x^m > \epsilon$. These last remarks are especially meaningful in light of the fact that the discontinuity of the limit function occurred at 1. We are led to suspect that a notion of "uniform" convergence, so-named because of the similar distinction between continuity and uniform continuity, might be in order.

DEFINITION A sequence $\{f_n\}_{n=1}^{\infty}$ of functions is said to *converge uniformly* on E if there is a function $f: E \to R$ such that for each $\epsilon > 0$, there is N such that for each positive integer n, $n \geq N$ implies that

$$|f_n(x) - f(x)| < \epsilon$$

for all $x \in E$. We may also express this by writing "the sequence $\{f_n\}_{n=1}^{\infty}$ converges uniformly to f on E."

Naturally, a series $\Sigma_{n=0}^{\infty} f_n$ of functions will converge pointwise or converge uniformly on E as the sequence $\{S_n\}_{n=0}^{\infty}$ of partial sums, defined by

$$S_n(x) = \sum_{k=0}^{n} f_k(x),$$

converges pointwise or uniformly on E.

The following theorem arises quite naturally after our previous experience with sequences. The proof is left to the reader as Exercise 4.

7.1 THEOREM A sequence of functions $\{f_n\}_{n=1}^{\infty}$ converges uniformly on E iff for each $\epsilon > 0$ there is a real number N such that for all positive integers m and n, $m \geq N$ and $n \geq N$ imply that

$$|f_n(x) - f_m(x)| < \epsilon$$

for all $x \in E$.

For infinite series of functions, there is a very convenient test for uniform convergence, called the Weierstrass M-test.

7.2 THEOREM (WEIERSTRASS M-TEST) Suppose $\{f_n\}_{n=1}^{\infty}$ is a sequence of functions defined on E and $\{M_n\}_{n=1}^{\infty}$ is a sequence of nonnegative real numbers such that

$$|f_n(x)| \leq M_n$$

for all $x \in E$ and all positive integers n. If $\Sigma_{n=1}^{\infty} M_n$ converges, then $\Sigma_{n=1}^{\infty} f_n$ converges uniformly on E.

Proof. Choose $\epsilon > 0$. Since $\Sigma_{n=1}^{\infty} M_n$ converges, there is a real number N such that for all positive integers m and n, $N \leq n \leq m$ implies that

$$\sum_{k=n}^{m} M_k < \epsilon.$$

(Recall that $M_n \geq 0$ for all positive integers n.) Thus, for all positive integers m and n, $N \leq n \leq m$ implies that

$$|S_m(x) - S_n(x)| = \left| \sum_{k=n+1}^{m} f_k(x) \right| \leq \sum_{k=n+1}^{m} M_k < \epsilon$$

for all $x \in E$. By Theorem 7.1, $\Sigma_{k=1}^{\infty} f_k$ converges uniformly on E.

We have exhibited remarkable confidence in the definition of uniform convergence by proving the last two theorems before having any concrete evidence that this new concept is really useful. The next few theorems serve to show that this confidence is not unfounded.

Suppose $\{f_n\}_{n=1}^{\infty}$ is a sequence of functions converging uniformly to f on E such that for each positive integer n, f_n is continuous at $x_0 \in E$. Choose $\epsilon > 0$. There is N such that for each positive integer n, $n \geq N$ implies that

$$|f_n(x) - f(x)| < \epsilon$$

for all $x \in E$. Since f_N is continuous at x_0, there is $\delta > 0$ such that $|x - x_0| < \delta$ and $x \in E$ imply that

$$|f_N(x) - f_N(x_0)| < \epsilon.$$

Hence, for $|x - x_0| < \delta$ and $x \in E$, we have

$$|f(x) - f(x_0)| \leq |f(x) - f_N(x)| + |f_N(x) - f_N(x_0)| + |f_N(x_0) - f(x_0)| < \epsilon + \epsilon + \epsilon = 3\epsilon.$$

It seems we have just proved a theorem. What is really necessary for this proof to yield the desired conclusion—that f is continuous at x_0?

First of all, it is sufficient that infinitely many members of the sequence $\{f_n\}_{n=1}^{\infty}$ be continuous at x_0. Second, uniform convergence on all of E is unnecessary. We now present a slight modification of our original definition of uniform convergence that will suffice for this proof.

DEFINITION Suppose $\{f_n\}_{n=1}^{\infty}$ is a sequence of functions converging pointwise to a function f on a set E. Then $\{f_n\}_{n=1}^{\infty}$ *converges uniformly at* $x_0 \in E$ if for each $\epsilon > 0$, there are positive real numbers N and δ such that for each positive integer n, $n \geq N$, $|x - x_0| < \delta$, and $x \in E$ imply that

$$|f_n(x) - f(x)| < \epsilon.$$

It is immediate that if $\{f_n\}_{n=1}^{\infty}$ converges uniformly to f on E, then $\{f_n\}_{n=1}^{\infty}$ converges uniformly at x for each $x \in E$. Conversely, if E is compact and $\{f_n\}_{n=1}^{\infty}$ converges uniformly at x for each $x \in E$, then $\{f_n\}_{n=1}^{\infty}$ converges uniformly on E. Can you prove this?

7.2 CONSEQUENCES OF UNIFORM CONVERGENCE

7.3 THEOREM Suppose $\{f_n\}_{n=1}^{\infty}$ converges pointwise to f on E, $x_0 \in E$, and infinitely many members of the sequence are continuous at x_0. If $\{f_n\}_{n=1}^{\infty}$ converges uniformly at x_0, then f is continuous at x_0.

Proof. Choose $\epsilon > 0$. There are positive real numbers N and δ such that for each positive integer $n \geq N$,

$$|f_n(x) - f(x)| < \frac{\epsilon}{3}$$

for all $x \in E$ such that $|x - x_0| < \delta$. There is a positive integer $n_0 \geq N$ such that f_{n_0} is continuous at x_0; hence there is $\delta_1 > 0$ such that $|x - x_0| < \delta_1$ and $x \in E$ imply that

$$|f_{n_0}(x) - f_{n_0}(x_0)| < \frac{\epsilon}{3}.$$

Let $\delta_2 = \min \{\delta, \delta_1\}$. Then, for $|x - x_0| < \delta_2$ and $x \in E$, we have

$$|f(x) - f(x_0)| \leq |f(x) - f_{n_0}(x)| + |f_{n_0}(x) - f_{n_0}(x_0)|$$
$$+ |f_{n_0}(x_0) - f(x_0)| < \frac{\epsilon}{3} + \frac{\epsilon}{3} + \frac{\epsilon}{3} = \epsilon.$$

Consequently, f is continuous at x_0.

In particular, it follows from Theorem 7.3 that if $\{f_n\}_{n=1}^{\infty}$ is a sequence of functions continuous on E and converging uniformly to f on E, then f is continuous. (Of course, it is implicit that $E = \text{dom } f$.)

The example presented earlier to convince the reader that the limit of a sequence of continuous functions need not be Riemann-integrable yielded a limit function that was unbounded. The added hypothesis of uniform convergence will serve to eliminate such occurrences.

7.4 THEOREM If $\{f_n\}_{n=1}^{\infty}$ is a sequence of functions converging uniformly to f on E and for each positive integer n, f_n is bounded on E, then f is bounded on E.

Proof. Let $\epsilon = 1$. There is a real number N such that for each positive integer n, $n \geq N$ implies that

$$|f_n(x) - f(x)| < 1$$

for all $x \in E$. There is M such that $|f_N(x)| \leq M$ for all $x \in E$. Hence, for all $x \in E$,

$$|f(x)| \leq |f(x) - f_N(x)| + |f_N(x)| \leq 1 + M.$$

One should now hope that, if $\{f_n\}_{n=1}^{\infty}$ converges uniformly to f on $[a, b]$ and for each positive integer n, f_n is Riemann-integrable on $[a, b]$, then f is Riemann-integrable on $[a, b]$. This and more is the content of the next theorem.

7.5 THEOREM Let $\{f_n\}_{n=1}^{\infty}$ be a sequence of functions, each Riemann-integrable on $[a, b]$, converging uniformly to f on $[a, b]$. Define

$$F_n(t) = \int_a^t f_n(x)\, dx$$

for each $t \in [a, b]$ and each positive integer n. Then f is Riemann-integrable on $[a, b]$, and $\{F_n\}_{n=1}^{\infty}$ converges uniformly on $[a, b]$ to the function F defined by

$$F(t) = \int_a^t f(x)\,dx$$

for each $t \in [a, b]$.

Proof. Each f_n is bounded on $[a, b]$ so, by Theorem 7.4, f is also bounded. We may suppose $a < b$. Choose $\epsilon > 0$. There is a real number N such that for each positive integer n, $n \geq N$ implies that

$$|f_n(x) - f(x)| < \frac{\epsilon}{3(b - a)}$$

for all $x \in [a, b]$. There is a partition P of $[a, b]$ such that

$$U(P, f_N) - L(P, f_N) < \frac{\epsilon}{3}.$$

For each $x \in [a, b]$,

$$f_N(x) - \frac{\epsilon}{3(b - a)} < f(x) < f_N(x) + \frac{\epsilon}{3(b - a)}.$$

Thus,

$$L(P, f_N) - \frac{\epsilon}{3} < L(P, f) \leq U(P, f) < U(P, f_N) + \frac{\epsilon}{3}.$$

Consequently,

$$U(P, f) - L(P, f) < U(P, f_N) - L(P, f_N) + \frac{2\epsilon}{3} < \epsilon.$$

Hence, f is Riemann-integrable on $[a, b]$.

Again, choose $\epsilon > 0$ and N such that for each positive integer n, $n \geq N$ implies that

$$|f_n(x) - f(x)| < \frac{\epsilon}{3(b - a)}$$

for all $x \in [a, b]$. Then for each positive integer n, $n \geq N$ implies that

$$\begin{aligned} |F_n(t) - F(t)| = \left| \int_a^t f_n(x)\,dx - \int_a^t f(x)\,dx \right| &\leq \int_a^t |f_n(x) - f(x)|\,dx \\ &\leq (t - a)\frac{\epsilon}{3(b - a)} \leq \frac{\epsilon}{3} < \epsilon \end{aligned}$$

for all $t \in [a, b]$. Therefore, $\{F_n\}_{n=1}^{\infty}$ converges uniformly to F on $[a,b]$.

Very soon we shall prove that the power series $\Sigma_{n=0}^{\infty} x^n$ converges uniformly on any closed interval $[a, b] \subset (-1, 1)$. (Do you want to try the proof now? Why not?) Assuming this fact, let us see what results we can obtain with the aid of Theorem 7.5. Choose any closed interval $[a, b] \subset (-1, 1)$. It is no surprise that

$$\sum_{n=0}^{\infty} x^n = \frac{1}{1 - x}$$

is Riemann-integrable on $[a, b]$. But Theorem 7.5 implies even more; it implies also that

$$\sum_{n=0}^{\infty} \int_0^x t^n \, dt = \sum_{n=0}^{\infty} \frac{x^{n+1}}{n + 1}$$

converges uniformly to $\int_0^x dt/(1 - t)$ on $[a, b]$. Recall that

$$\int_0^x \frac{dt}{1 - t} = -\log (1 - x),$$

so we have a power series $\Sigma_{n=0}^{\infty} \dfrac{x^{n+1}}{n + 1}$ that converges to $-\log (1 - x)$ on $(-1, 1)$ and converges uniformly on any closed subinterval of $(-1, 1)$.

The concept of uniform convergence does not allow affirmative answers to all the questions posed at the beginning of this chapter. The sequence $\{k_n\}_{n=1}^{\infty}$, defined on $[-1, 1]$ by

$$k_n(x) = \frac{x}{1 + nx^2}$$

for each positive integer n and each $x \in [-1, 1]$, converges uniformly to the function that is zero everywhere on $[-1, 1]$. However, $\{k_n'(0)\}_{n=1}^{\infty}$ converges to 1. Consequently, one must have some additional hypotheses to gain the type of result we have in mind.

7.6 THEOREM Suppose $\{f_n\}_{n=1}^{\infty}$ is a sequence of functions, each of which is differentiable on $[a, b]$. Suppose further that for some $x_0 \in [a, b]$,

$$\{f_n(x_0)\}_{n=1}^{\infty}$$

converges and that $\{f_n'\}_{n=1}^{\infty}$ converges uniformly to g on $[a, b]$. Then

1. $\{f_n\}_{n=1}^{\infty}$ converges uniformly on $[a, b]$ to a function f;
2. f is differentiable on $[a, b]$ and $f'(x) = g(x)$ for all $x \in [a, b]$.

Proof. Assume $b > a$ and choose $\epsilon > 0$. There is a real number N such that for all positive integers m and n, $m \geqslant N$ and $n \geqslant N$ imply that

$$|f_n(x_0) - f_m(x_0)| < \frac{\epsilon}{2}$$

and for all $t \in [a, b]$,

$$|f_n'(t) - f_m'(t)| < \frac{\epsilon}{2(b - a)}.$$

For each pair of positive integers m and n, $f_n - f_m$ is differentiable on $[a, b]$; hence, for each $u \in [a, b]$ and $v \in [a, b]$, there is $w \in [a, b]$ such that

$$[f_n(u) - f_m(u)] - [f_n(v) - f_m(v)] = [f_n'(w) - f_m'(w)](u - v).$$

Therefore, for all positive integers m and n, $m \geqslant N$ and $n \geqslant N$ imply that

$$\begin{aligned}|f_n(x) - f_m(x)| &\leqslant |f_n(x) - f_m(x) - f_n(x_0) + f_m(x_0)| \\ &\quad + |f_n(x_0) - f_m(x_0)| \\ &< \frac{\epsilon|x - x_0|}{2(b - a)} + \frac{\epsilon}{2} \leqslant \frac{\epsilon}{2} + \frac{\epsilon}{2} = \epsilon\end{aligned}$$

for all $x \in [a, b]$. Consequently $\{f_n\}_{n=1}^{\infty}$ converges uniformly on $[a, b]$. Define

$$f(x) = \lim_{n\to\infty} f_n(x)$$

for each $x \in [a, b]$.

Choose $x \in [a, b]$ and for each positive integer n, define F_n: $[a, b] \to R$ by

$$F_n(y) = \frac{f_n(x) - f_n(y)}{x - y}$$

for all $y \in [a, b] \setminus \{x\}$ and $F_n(x) = f_n'(x)$. Define F: $[a, b] \to R$ by

$$F(y) = \frac{f(x) - f(y)}{x - y}$$

for all $y \in [a, b] \setminus \{x\}$ and $F(x) = g(x)$. The function f will be differentiable at x with $f'(x) = g(x)$ if F is continuous at x. For each positive integer n, F_n is continuous at x by the differentiability of f_n at x. If we can show that $\{F_n\}_{n=1}^{\infty}$ converges uniformly to F on $[a, b]$, then F will be continuous at x, and hence f will be differentiable at x with $f'(x) = g(x)$. Choose $\epsilon > 0$. As before, there is a real number N such that for all positive integers m and n, $m \geqslant N$ and $n \geqslant N$ imply that for all $t \in [a, b]$,

$$|f_n'(t) - f_m'(t)| < \frac{\epsilon}{2}.$$

Then, for all positive integers m and n, $m \geqslant N$ and $n \geqslant N$ imply that

$$\begin{aligned}|F_n(y) - F_m(y)| &= \left|\frac{f_n(x) - f_n(y)}{x - y} - \frac{f_m(x) - f_m(y)}{x - y}\right| \\ &= \left|\frac{1}{x - y}\right| \cdot |f_n(x) - f_n(y) - f_m(x) + f_m(y)| \\ &\leqslant \frac{1}{|x - y|} \cdot \frac{\epsilon}{2} \cdot |x - y| < \epsilon\end{aligned}$$

for all $y \in [a, b] \setminus \{x\}$, and

$$|F_n(x) - F_m(x)| = |f_n'(x) - f_m'(x)| < \frac{\epsilon}{2} < \epsilon.$$

Thus, by Theorem 7.1, $\{F_n\}_{n=1}^{\infty}$ converges uniformly on $[a, b]$, and it is clear that $\{F_n\}_{n=1}^{\infty}$ converges to F. This concludes the proof.

7.3 UNIFORM CONVERGENCE OF POWER SERIES

Before examining in more detail the results of Theorems 7.3, 7.5, and 7.6, we return to consider the convergence of power series. The Weierstrass M-test suggests the following theorem.

7.7 THEOREM Let $\Sigma_{n=0}^{\infty} a_n x^n$ be a power series that converges for $-r < x < r$, $r > 0$. Then $\Sigma_{n=0}^{\infty} a_n x^n$ converges uniformly on $-t \leq x \leq t$ for each $0 < t < r$.

Proof. If $0 < t < r$, then $\Sigma_{n=0}^{\infty} a_n t^n$ converges absolutely. For $-t \leq x \leq t$,

$$|a_n x^n| \leq |a_n| t^n.$$

Thus, by the Weierstrass M-test, $\Sigma_{n=0}^{\infty} a_n x^n$ converges uniformly on $[-t, t]$.

Consider a power series $\Sigma_{n=0}^{\infty} a_n x^n$. To apply the results of Theorem 7.6, one must consider the question of convergence of the series

$$\sum_{n=0}^{\infty} n a_n x^{n-1}.$$

It is not at all clear from casual observation that this series should converge for any $x \neq 0$. Even though $\Sigma_{n=0}^{\infty} a_n x^n$ converges, it seems reasonable that $\{n a_n x^{n-1}\}_{n=0}^{\infty}$ might not even converge to zero. Indeed, if $a_n = \frac{1}{n}$ and $x = -1$, this is the case. We have, from Chapter 6, a result that allows the direct determination of the interval of convergence of $\Sigma_{n=0}^{\infty} n a_n x^{n-1}$. Perhaps considerations of this type will give us the proper perspective.

7.8 LEMMA If $\{b_n\}_{n=1}^{\infty}$ is a bounded sequence of real numbers, then $\{\sqrt[n]{n} b_n\}_{n=1}^{\infty}$ is a bounded sequence of real numbers and

$$\limsup_{n \to \infty} b_n = \limsup_{n \to \infty} \sqrt[n]{n} b_n.$$

Proof. Let

$$b = \limsup_{n \to \infty} b_n.$$

In Chapter 1, we proved that $\{\sqrt[n]{n}\}_{n=1}^{\infty}$ converges to 1. Thus, $\{\sqrt[n]{n}\, b_n\}_{n=1}^{\infty}$ must be bounded. Let

$$c = \limsup_{n \to \infty} \sqrt[n]{n}\, b_n.$$

Let $\{b_{n_k}\}_{k=1}^{\infty}$ be a subsequence of $\{b_n\}_{n=1}^{\infty}$ converging to b. Then

$$\{\sqrt[n_k]{n_k}\,b_k\}_{k=1}^{\infty}$$

also converges to b and $b \leqslant c$. Suppose

$$\{\sqrt[m_j]{m_j}\,b_{m_j}\}_{j=1}^{\infty}$$

converges to c. Since $\{\sqrt[m_j]{m_j}\}_{j=1}^{\infty}$ converges to $1 \neq 0$, we conclude that

$$\{b_{m_j}\}_{j=1}^{\infty} = \left\{\sqrt[m_j]{m_j}\, b_{m_j} \cdot \frac{1}{\sqrt[m_j]{m_j}}\right\}_{j=1}^{\infty}$$

converges to c. Consequently, $c \leqslant b$. We have shown that $b \leqslant c$ and $c \leqslant b$, so $c = b$.

Lemma 7.8 and Theorem 6.18 yield the fact that

$$\sum_{n=0}^{\infty} a_n x^n \quad \text{and} \quad \sum_{n=0}^{\infty} n a_n x^n$$

have the same radius of convergence. It is easy to see that this implies that

$$\sum_{n=0}^{\infty} a_n x^n \quad \text{and} \quad \sum_{n=1}^{\infty} n a_n x^{n-1}$$

also have the same radius of convergence.

7.9 THEOREM Suppose $\Sigma_{n=0}^{\infty} a_n x^n$ converges to f on $(-r, r)$ with $r > 0$. Then

1. For each $0 < t < r$, $\Sigma_{n=0}^{\infty} a_n x^n$ converges uniformly on $[-t, t]$.
2. f is n-times differentiable on $(-r, r)$ for each positive integer n.
3. For each $0 < t < r$ and each positive integer m,

$$\sum_{n=m}^{\infty} n(n-1)\cdots(n-m+1)a_n x^{n-m}$$

converges uniformly to $f^{(m)}(x)$ on $[-t, t]$.

4. $f^{(n)}(0) = n!a_n$. [We make the convention that $f^{(0)}(x) = f(x)$ for all $x \in (-r, r)$.]

Proof. Part 1 of this theorem, a repetition of Theorem 7.7, is included only for the sake of completeness.

The proof of parts 2, 3, and 4 shall be by induction. Choose any $x_0 \in (-r, r)$. There is $0 < t < r$ such that $|x_0| < t$. Since $\Sigma_{n=0}^{\infty} a_n x^n$ converges on $(-r, r)$, Lemma 7.8 shows that

$$\sum_{n=1}^{\infty} n a_n x^{n-1}$$

converges on $(-r, r)$. By Theorem 7.7,

$$\sum_{n=1}^{\infty} na_n x^{n-1}$$

converges uniformly on $[-t, t]$. Define g: $[-t, t] \to R$ by $g(x) = f(x)$ for all $x \in [-t, t]$. Theorem 7.6 guarantees that g is differentiable on $[-t, t]$ and

$$g'(x) = \sum_{n=1}^{\infty} na_n x^{n-1}$$

for all $x \in [-t, t]$. Thus, since $x_0 \in (-t, t)$, f is differentiable at x_0 and

$$f'(x_0) = g'(x_0) = \sum_{n=1}^{\infty} na_n x^{n-1}.$$

Since x_0 was any point in $(-r, r)$, we have f differentiable on $(-r, r)$,

$$f'(x) = \sum_{n=1}^{\infty} na_n x^{n-1}$$

for all $x \in (-r, r)$, and $\sum_{n=1}^{\infty} na_n x^{n-1}$ converges uniformly to $f'(x)$ on any closed subinterval of $(-r, r)$. In particular, $f'(0) = a_1$.

Assume the theorem holds for $n = k$. Thus, f is k-times differentiable on $(-r, r)$; for each $0 < t < r$,

$$\sum_{n=k}^{\infty} n(n - 1) \cdots (n - k + 1)a_n x^{n-k}$$

converges uniformly to $f^{(k)}(x)$ on $[-t, t]$; and

$$f^{(k)}(0) = k!a_k.$$

Define h: $(-r, r) \to R$ by

$$h(x) = f^{(k)}(x)$$

for each $x \in (-r, r)$ and define

$$b_n = (n + k)(n + k - 1) \cdots (n + 1)a_{n+k}$$

for each nonnegative integer n. Then

$$h(x) = \sum_{n=0}^{\infty} b_n x^n$$

for all $x \in (-r, r)$. By the results in the preceding paragraph, h is differentiable on $(-r, r)$; for each $0 < t < r$,

$$\sum_{n=1}^{\infty} nb_n x^{n-1}$$

converges uniformly to $h'(x)$ on $[-t, t]$; and $h'(0) = b_1$. Thus, $f^{(k)}$ is differentiable on $(-r, r)$; for each $0 < t < r$,

$$\sum_{n=1}^{\infty} nb_n x^{n-1} = \sum_{n=k+1}^{\infty} n(n-1)\cdots(n-k)a_n x^{n-k-1}$$

converges uniformly to $f^{(k+1)}(x)$ on $[-t, t]$; and

$$f^{(k+1)}(0) = b_1 = (k+1)!a_{k+1}.$$

This completes the induction.

Theorem 7.9 tells us that, inside the interval of convergence (not necessarily at the endpoints), a power series may be differentiated term by term, and the resulting power series will converge to the derivative of the limit function of the original series. Theorem 7.5, coupled with Theorem 7.7, justifies the formula

$$\int_a^b \sum_{n=0}^{\infty} a_n x^n \, dx = \sum_{n=0}^{\infty} a_n \int_a^b x^n \, dx$$

for $-r < a \leqslant b < r$ if $\Sigma_{n=0}^{\infty} a_n x^n$ converges for $-r < x < r$.

Recall that $\Sigma_{n=0}^{\infty} x_n$ converges to $\dfrac{1}{1-x}$ for $-1 < x < 1$. Hence, by Theorem 7.9,

$$\sum_{n=1}^{\infty} nx^{n-1}$$

converges to $\dfrac{1}{(1-x)^2}$ and

$$\sum_{n=2}^{\infty} n(n-1)x^{n-2}$$

converges to $\dfrac{2}{(1-x)^3}$ for $-1 < x < 1$.

If $f: E \to R$ with E a neighborhood of 0, and if f has derivatives of all orders at zero, one can always consider the power series

$$\sum_{n=0}^{\infty} \frac{f^{(n)}(0)}{n!} x^n.$$

This series converges to $f(0)$ at zero, but it is not necessarily true that there is $r > 0$ such that

$$f(x) = \sum_{n=0}^{\infty} \frac{f^{(n)}(0)}{n!} x^n$$

for $-r < x < r$. The classic example is the function $f: R \to R$ defined by

$$f(x) = e^{-1/x^2}$$

for $x \neq 0$ and $f(0) = 0$. In this case, $f^{(n)}(0)$ exists and equals 0 for each positive integer n, but then

$$\sum_{n=0}^{\infty} \frac{f^n(0)}{n!} x^n = 0$$

for all x, and $f(x) = 0$ only for $x = 0$.

Consider the differential equation

$$x^2 f''(x) + x(1 - 2x)f'(x) + (x^2 - x)f(x) = 0.$$

In Chapter 1, we proved that $\{\sqrt[n]{n}\}_{n=1}^{\infty}$ converges to 1 by assuming convergence and showing that then the limit must be 1; and then we proved that the sequence converged to 1. A similar device will be used here. Suppose there is a function f satisfying this differential equation, and suppose that

$$f(x) = \sum_{n=0}^{\infty} a_n x^n$$

for $-r < x < r$ with $r > 0$. Then for $-r < x < r$,

$$f'(x) = \sum_{n=1}^{\infty} n a_n x^{n-1}$$

and

$$f''(x) = \sum_{n=2}^{\infty} n(n - 1)a_n x^{n-2}.$$

Consequently, for $-r < x < r$,

$$\begin{aligned}
0 &= x^2 f''(x) + x(1 - 2x)f'(x) + (x^2 - x)f(x) \\
&= x^2 \sum_{n=2}^{\infty} n(n - 1)a_n x^{n-2} + x(1 - 2x) \sum_{n=1}^{\infty} n a_n x^{n-1} \\
&\quad + (x^2 - x) \sum_{n=0}^{\infty} a_n x^n \\
&= \sum_{n=2}^{\infty} n(n - 1)a_n x^n + \sum_{n=1}^{\infty} n a_n x^n - 2 \sum_{n=1}^{\infty} n a_n x^{n+1} + \sum_{n=0}^{\infty} a_n x^{n+2} \\
&\quad - \sum_{n=0}^{\infty} a_n x^{n+1} = (a_1 - a_0)x + \sum_{n=2}^{\infty} [n^2 a_n - (2n - 1)a_{n-1} + a_{n-2}]x^n.
\end{aligned}$$

This equation will be satisfied if $\{a_n\}_{n=0}^{\infty}$ is chosen so that $a_1 = a_0$ and for each integer $n \geqslant 2$,

$$n^2 a_n - (2n - 1)a_{n-1} + a_{n-2} = 0.$$

Thus, we may choose $a_0 = a_1$ and, for each integer $n \geqslant 2$,

$$a_n = \frac{(2n - 1)a_{n-1} - a_{n-2}}{n^2}.$$

Several questions must be considered at this time. First of all, if a sequence $\{a_n\}_{n=0}^{\infty}$ is chosen so that

$$a_0 = a_1 \quad \text{and} \quad a_n = \frac{(2n - 1)a_{n-1} - a_{n-2}}{n^2}$$

for each integer $n \geqslant 2$, is it true that $\sum_{n=0}^{\infty} a_n x^n$ will converge for some $x \neq 0$? Secondly, there may be other methods of choosing the sequence $\{a_n\}_{n=0}^{\infty}$ to satisfy the equation above. More to the second point, does there exist a sequence $\{b_n\}_{n=0}^{\infty}$ such that $\sum_{n=0}^{\infty} b_n x^n$ converges to zero on an interval of positive length with $b_n \neq 0$ for some integers n? We answer the second question first.

7.10 THEOREM Suppose $\{a_n\}_{n=0}^{\infty}$ and $\{b_n\}_{n=0}^{\infty}$ are two sequences of real numbers, $r > 0$, and for all $x \in (-r, r)$,

$$\sum_{n=0}^{\infty} a_n x^n = \sum_{n=0}^{\infty} b_n x^n.$$

Then for each integer $n \geqslant 0$, $a_n = b_n$.

Proof. Define $f\colon (-r, r) \to R$ by

$$f(x) = \sum_{n=0}^{\infty} a_n x^n = \sum_{n=0}^{\infty} b_n x^n$$

for each $x \in (-r, r)$. Then, by Theorem 7.9, f has derivatives of all orders on $(-r, r)$, and for each integer $n \geqslant 0$,

$$f^{(n)}(0) = n!a_n = n!b_n.$$

Thus, for each integer $n \geqslant 0$, $a_n = b_n$.

In particular, Theorem 7.10 allows one to conclude that if $\sum_{n=0}^{\infty} a_n x^n$ converges to zero for all $x \in (-r, r)$ with $r > 0$, then $a_n = 0$ for each integer $n \geqslant 0$. Therefore, any solution to the differential equation

$$x^2 f''(x) + x(1 - 2x)f'(x) + (x^2 - x)f(x) = 0$$

that is expressible as a power series $\sum_{n=0}^{\infty} a_n x^n$, convergent on an interval of positive radius, must satisfy

$$a_1 = a_0 \quad \text{and} \quad a_n = \frac{(2n - 1)a_{n-1} - a_{n-2}}{n^2}.$$

The question converning the convergence of such a power series is still unanswered. To simplify matters, choose $a_0 = a_1 = 1$. Then $a_2 = \frac{1}{2}$, $a_3 = \frac{1}{6}$, $\ldots$, $a_n = \frac{1}{n!}$, ... The solution then is

$$f(x) = \sum_{n=0}^{\infty} \frac{x^n}{n!} = e^x,$$

and this solution is valid for all x since $\sum_{n=0}^{\infty} \frac{x^n}{n!}$ converges for all x. In this case, it is trivial to check that this is a solution, since the function $f(x) = e^x$ is well known to us.

Now let us apply the result of Theorem 7.5 to power series.

7.11 THEOREM Suppose $\sum_{n=0}^{\infty} a_n(x - a)^n$ converges to f for $-r < x - a < r$ with $r > 0$. Then

$$\sum_{n=0}^{\infty} \frac{a_n}{n + 1}(x - a)^{n+1}$$

converges to

$$\int_a^x f(t)\, dt$$

for $-r < x - a < r$.

Proof. This is an easy application of Theorems 7.9 and 7.5 using the fact that

$$\int_a^x a_n(t - a)^n\, dt = \frac{a_n}{n + 1}(x - a)^{n+1}.$$

Theorems 7.9, 7.10, and 7.11 allow us to find the power series for many functions without resorting to Taylor's Theorem.

■ **Example 7.6** We seek a power series for $f(x) = \text{Arctan } x$. Assuming the usual facts about the trigonometric functions and their derivatives, we find that $f'(x) = \frac{1}{1 + x^2}$. To find successive derivatives of f will require applications of the quotient rule, and it may be difficult to find a nice formula for $f^{(n)}(x)$. Rather, we will use Theorem 7.11. We know that

$$\text{Arctan } x = \int_0^x \frac{1}{1 + t^2}\, dt,$$

so we will find a power series for $\frac{1}{1 + x^2}$ and apply Theorem 7.11 to find the series for Arctan x. From Chapter 6,

$$\frac{1}{1 - x} = \sum_{n=0}^{\infty} x^n \quad \text{for } -1 < x < 1.$$

Thus

$$\frac{1}{1+x^2} = \frac{1}{1-(-x^2)} = \sum_{n=0}^{\infty} (-x^2)^n = \sum_{n=0}^{\infty} (-1)^n x^{2n}$$

for $-1 < x^2 < 1$ or $-1 < x < 1$. Theorem 7.11 allows us to find the series for Arctan x by integrating the series just obtained term by term:

$$\text{Arctan } x = \sum_{n=0}^{\infty} \frac{(-1)^n}{2n+1} x^{2n+1} \quad \text{for } -1 < x < 1. \quad \blacksquare$$

■ **Example 7.7** Power series may also be useful in evaluating certain limits. Consider the next example.

Define $f(x) = \dfrac{e^x - 1}{x}$ for all $x \neq 0$. We seek to determine whether f has a limit at zero and, if so, what the limit is. We may rewrite the numerator as a power series:

$$e^x - 1 = \left(1 + x + \frac{x^2}{2!} + \cdots\right) - 1 = x + \frac{x^2}{2!} + \cdots;$$

so, for $x \neq 0$,

$$f(x) = \frac{e^x - 1}{x} = \frac{x + \dfrac{x^2}{2!} + \cdots}{x} = 1 + \frac{x}{2!} + \frac{x^2}{3!} + \cdots.$$

We know that the series $1 + \dfrac{x}{2!} + \dfrac{x^2}{3!} + \cdots$ converges for all x to a function g that is continuous for all x by Theorem 7.9. Also, $f(x) = g(x)$ for all $x \neq 0$, hence f has a limit at 0 because g does, and $\lim_{x\to 0} f(x) = g(0) = 1$. ■

EXERCISES

7.1 POINTWISE AND UNIFORM CONVERGENCE

1. Let f: $(a, b) \to R$ be differentiable. Construct a sequence of continuous functions that converges pointwise to f' on (a, b).
2. Let $\{r_n\}_{n=1}^{\infty}$ be a sequence containing each rational number in $[0, 1]$ exactly once. Define, for each positive integer n,

$$\begin{aligned} f_n(x) &= 0 \quad \text{if } x \text{ is irrational,} \\ f_n(r_i) &= 0 \quad \text{if } i > n, \text{ and} \\ f_n(r_i) &= 1 \quad \text{if } i \leq n. \end{aligned}$$

   Show that $\{f_n\}_{n=1}^{\infty}$ converges pointwise on $[0, 1]$ to a function f that is not Riemann-integrable.

3. Prove that the sequence $\{k_n\}_{n=1}^{\infty}$, defined by

$$k_n(x) = \frac{x}{1 + nx^2}$$

for all $x \in R$ and each positive integer n, converges uniformly on R.

4. Prove Theorem 7.1.

5. Suppose that both $\{f_n\}_{n=1}^{\infty}$ and $\{g_n\}_{n=1}^{\infty}$ converge uniformly on E. Prove that $\{f_n + g_n\}_{n=1}^{\infty}$ converges uniformly on E. What can one prove about the uniform convergence of $\{f_n g_n\}_{n=1}^{\infty}$?

6. Let $f\colon R \to R$ be uniformly continuous, and for each positive integer n and each $x \in R$, define

$$f_n(x) = f\left(x + \frac{1}{n}\right).$$

Prove that $\{f_n\}_{n=1}^{\infty}$ converges uniformly to f on R.

7. Assume E is compact and $\{f_n\}_{n=1}^{\infty}$ is a sequence of functions that converges uniformly on each point of E. Prove that $\{f_n\}_{n=1}^{\infty}$ converges uniformly on E.

7.2 CONSEQUENCES OF UNIFORM CONVERGENCE

8. If $A \subset E \subset R$, then A is *dense* in E if $E = \overline{A} \cap E$. Assume $\{f_n\}_{n=1}^{\infty}$ is a sequence of functions continuous on E and converging uniformly on a set A dense in E. Prove that $\{f_n\}_{n=1}^{\infty}$ converges uniformly on E.

9. For each positive integer n, define $h_n\colon [0, 1] \to R$ by

$$h_n(x) = \frac{1}{x} \quad \text{for } \frac{1}{n} \leq x \leq 1 \text{ and}$$

$$h_n(x) = n^2 x \quad \text{for } 0 \leq x < \frac{1}{n}.$$

Show that $\{h_n\}_{n=1}^{\infty}$ does not converge uniformly on $[0, 1]$ but does converge uniformly on $[a, 1]$ for each $0 < a < 1$.

10. Search the literature for an example of a continuous, nowhere-differentiable function. Such an example will probably be more meaningful in light of the contents of this chapter.

11. If $\{f_n\}_{n=1}^{\infty}$ is a sequence of continuous functions that converges pointwise to f on $[a, b]$, then there is at least one point $x_0 \in (a, b)$ such that $\{f_n\}_{n=1}^{\infty}$ converges uniformly at x_0. Search the literature for a proof of this theorem.

12. Suppose E is compact and $\{f_n\}_{n=1}^{\infty}$ is a sequence of continuous functions defined on E that converges pointwise to a function f, also continuous on E. If, for each positive integer n,

$$f_n(x) \leq f_{n+1}(x)$$

for all $x \in E$, prove that $\{f_n\}_{n=1}^{\infty}$ converges uniformly to f.

7.3 UNIFORM CONVERGENCE OF POWER SERIES

13. Show that the power series $\sum_{n=0}^{\infty} \frac{x^n}{n!}$ converges uniformly on $[-a, a]$ for any real number a.

14. Find the power series for the function $f(x) = \dfrac{1}{2 - 3x}$ and find the interval in which it converges to f. You may want to write f in the form $\dfrac{a}{1 - bx}$ and use a geometric series.

15. Find the power series for the function $\log(1 - x)$ and find the interval in which it converges to $\log(1 - x)$.

16. Define $f(x) = \dfrac{2 - x^2 - 2\cos x}{x^4}$ for $x \neq 0$. Determine whether f has a limit at zero, and if so, find the limit.

17. Find the first three nonzero terms of the power series for Arcsin x. In what interval does it converge to Arcsin x?

18. Define $f(x) = \dfrac{\sin x - x}{x^2 \operatorname{Arcsin} x}$ for $x \neq 0$. Determine whether f has a limit at zero, and if so, find the limit.

PROJECT

The purpose of this project is to prove Abel's Theorem concerning convergence of power series at endpoints of the interval of convergence. We know that given a power series, the interval of convergence is R, $\{0\}$, or an interval of positive length. If the interval of convergence is $(-a, a)$, we also know that the series converges uniformly on $[-s, s]$ for an $0 < s < a$. However, the question of convergence at the endpoints is unanswered. Indeed, we have seen enough examples to expect that the series may converge at all, none, or only one of the endpoints. Suppose that the series in question converges at a. Does the series converge uniformly at a? We haven't addressed that question as yet, and thereby hangs the tale. We will state Abel's Theorem in the most general way, and then prove only a special case. You are invited to adapt the proof to the general situation.

ABEL'S THEOREM If $\Sigma_{n=0}^{\infty} a_n(x - x_0)^n$ converges for $x_0 - R < x < x_0 + R$ and if $\Sigma_{n=0}^{\infty} a_n R^n$ converges, then the series converges uniformly on $[x_0, x_0 + R]$. Similarly, if $\Sigma_{n=0}^{\infty} a_n(-R)^n$ converges, the series converges uniformly on $[x_0 - R, x_0]$.

The version of this theorem that we will prove is the following:

THEOREM If $\Sigma_{n=0}^{\infty} a_n$ converges, then $\Sigma_{n=0}^{\infty} a_n x^n$ converges uniformly on $[0, 1]$.

The word *we* used above is a euphemism for "We will give hints and direction and you will write the proof." So get out your pencil and paper and be ready to go to work.

1. Assume $\Sigma_{n=0}^{\infty} a_n$ converges and $0 \leq x < 1$. Let $A_n = \Sigma_{k=n}^{\infty} a_k$. Show that

$$\sum_{k=n}^{\infty} a_k x^k = A_n x^n + x^n(x - 1) \sum_{k=n}^{\infty} A_{k+1} x^{k-n}.$$

2. Choose $\epsilon > 0$. There is N such that $|A_n| < \frac{\epsilon}{2}$ for all $n \geq N$. Then for $n \geq N$, show that

$$\left| \sum_{k=n}^{\infty} a_k x^k \right| < \epsilon \quad \text{if } 0 \leq x < 1.$$

For $x = 1$, the inequality is just $|A_n| < \epsilon$. Thus, the series converges uniformly on $[0, 1]$. [You may want to use the fact that $\Sigma_{m=0}^{\infty} x^m = \frac{1}{1 - x}$.]

As a result of our special case of Abel's Theorem, we can conclude something about the behavior of the function defined by the power series if the series converges at an endpoint. We invite you to state and prove a more general theorem based on Abel's Theorem.

THEOREM If f is defined by $f(x) = \Sigma_{n=0}^{\infty} a_n x^n$ for $-1 < x < 1$, $\Sigma_{n=0}^{\infty} a_n$ converges, and $f(1)$ is defined to be $\Sigma_{n=0}^{\infty} a_n$, then f is continuous at 1.

3. Supply a proof for the theorem above. (You shouldn't need any hints on this one.)

As a result of all of this, we can now supply a proof of Theorem 6.15.

THEOREM Suppose $\Sigma_{n=0}^{\infty} a_n$ converges to A and $\Sigma_{n=0}^{\infty} b_n$ converges to B, and let $\Sigma_{n=0}^{\infty} c_n$ be the Cauchy product of $\Sigma_{n=0}^{\infty} a_n$ and $\Sigma_{n=0}^{\infty} b_n$. If $\Sigma_{n=0}^{\infty} c_n$ converges to C, then $C = AB$.

4. Supply a proof of the theorem above. As a suggestion, define two functions $f(x) = \Sigma_{n=0}^{\infty} a_n x^n$ and $g(x) = \Sigma_{n=0}^{\infty} b_n x^n$, and use some of the results in this project as well as Theorem 6.14.

Index